Test Bank

INTRODUCTORY CHEMISTRY: A FOUNDATION

INTRODUCTORY CHEMISTRY

BASIC CHEMISTRY

FIFTH EDITION

Steven S. Zumdahl

Donald J. DeCoste

HOUGHTON MIFFLIN COMPANY   BOSTON   NEW YORK

Editor-in-Chief: Charles Hartford
Executive Editor: Richard Stratton
Senior Development Editor: Bess Deck
Editorial Assistant: Rosemary Mack
Senior Project Editor: Cathy Brooks
Manufacturing Assistant: Karmen Chong
Senior Marketing Manager: Katherine Greig
Marketing Associate: Alexandra Shaw

Printed in the U.S.A.

ISBN: 0-618-30535-1

123456789 – EB – 07 06 05 04 03

Contents

To the Instructor

The questions in each chapter of this *Test Bank* are arranged according to the section in the textbook. The test items consist of multiple-choice, true-false, short-answer, matching, and completion (fill-in-the-blank) questions. Altogether, the *Test Bank* provides more than 1600 questions for use in tests, quizzes, and examinations. Computerized versions of the *Test Bank* are available for MS-DOS, Macintosh, and Windows.

Because of the large number of questions in the chapter, Chapter 5 has been divided into two parts as follows:

Chapter 5a: Sections 5.1 through 5.6
Chapter 5b: Section 5.7

In Chapter 5a, the questions appearing under the heading "Sections 5.2-5.6" combine material from all these sections. To use Chapter 5a or Chapter 5b on the computerized version, simply type in 5a or 5b at the chapter prompt. When using the random select feature on the computerized version of the *Test Bank,* key in the number of questions desired. (Optional: You also may key in a letter to choose a question type. Select M for multiple-choice, T for true-false, S for short-answer, K for matching, or C for completion.) However, you must leave the difficulty level blank.

For more information about the computerized version of this *Test Bank,* please call College Marketing Services at (800) 235-3565.

CHAPTER 1
Chemistry: An Introduction

1. Define the following terms:
   a. science
   b. chemistry

   ANSWER:

2. Define the following terms:
   a. scientific method
   b. natural law
   c. hypothesis
   d. theory

   ANSWER:

3. Discuss how the scientific method is useful in solving problems outside science.

   ANSWER:

4. A _____ is a summary of observed behavior and a _____ is an explanation of behavior.
   a. law, measurement
   b. theory, scientific method
   c. theory, law
   d. law, therory
   e. hypothesis, theory

   ANSWER: d. law, theory

CHAPTER 2

Measurements and Calculations

1. Express 1870000 in scientific notation.
   a. 5.49×10^{-8}
   b. 1.87×10^{-6}
   c. 1.87×10^{6}
   d. 187×10^{6}
   e. 187×10^{4}

   ANSWER: c. 1.87×10^{6}

2. Express 30501000 in scientific notation.
   a. 3×10^{7}
   b. 3.0501×10^{7}
   c. 351×10^{7}
   d. 30501×10^{3}
   e. 305010×10^{7}

   8/03

   ANSWER: b. 3.0501×10^{7}

3. Write 8,319 in standard scientific notation.
   a. 8,319
   b. 8.319×10^{-3}
   c. 831.9×10^{1}
   d. 8.319×1000
   e. 8.319×10^{3}

   ANSWER: e. 8.319×10^{3}

4. The number 0.00003007 expressed in exponential notation is
   a. 3.007×10^{-5}
   b. 3.7×10^{-5}
   c. 3.007×10^{5}
   d. 3.007×10^{-4}
   e. 3.007

   8/03

   ANSWER: a. 3.007×10^{-5}

2

5. The number 0.001 expressed in exponential notation is
    a. 1×10^3
    b. 1×10^4
    c. 1×10^{-3}
    d. 1×10^{-4}
    e. none of these

    ANSWER: c. 1×10^{-3}

6. The number 0.00231 expressed in exponential notation is
    a. 2.31×10^3
    b. 2.31×10^{-2}
    c. 231×10^3
    d. 2.31×10^2
    e. 2.31×10^{-3}

    ANSWER: e. 2.31×10^{-3}

7. The number 0.005802 expressed in scientific notation is
    a. 5.82×10^3
    b. 5.802×10^3
    c. 5.82×10^{-3}
    d. 5.802×10^{-3}
    e. 5802×10^{-6}

    ANSWER: d. 5.802×10^{-3}

8. The number 200,000 expressed in scientific notation is
    a. 2.0×10^5
    b. 2.0×10^{-5}
    c. 20×10^4
    d. 200×10^3
    e. 2×10^5

    ANSWER: e. 2×10^5

9. Express the following in scientific notation.
    a. 0.00832
    b. 178,428

    ANSWER:
    a. 8.32×10^{-3}
    b. 1.78428×10^5

10. Express the following numbers in scientific notation.
    a.   6020
    b.   0.00304

    ANSWER:
    a. 6.020×10^3
    b. 3.004×10^{-3}

11. One kilogram contains this many grams.
    a.   1000
    b.   100
    c.   10
    d.   1/100
    e.   1/1000

    ANSWER: a.   1000

12. How many milliliters are in 0.020 L?
    a.   0.20 mL
    b.   2.0 mL
    c.   5.0 mL
    d.   20. mL
    e.   200 mL

    ANSWER: d.   20. mL

13. The measurement 4.3×10^3 g also could be written as
    a.   4.3 g
    b.   4.3 mg
    c.   4.3 pg
    d.   4.3 kg
    e.   4.3 dg

    ANSWER: d.   4.3 kg

14. How many millimeters are in 251 centimeters?
    a. 2.51×10^2 mm
    b. 2.51×10^1 mm
    c. 2.51×10^3 mm
    d.   2.51 mm
    e. 2.51×10^{-2} mm

    ANSWER: c. 2.51×10^3 mm

15. Convert: 1 mm = _____ km.

    ANSWER: 1×10^{-6} km

16. Convert: 12.4 L = _____ mL.

    ANSWER: 1.24×10^4 mL

17.   Convert: 314 g = _____ kg.

ANSWER: 0.314 kg

18.   Convert: 50.0 cm = _____ m.

ANSWER: 0.500 m

19.   One millisecond is equal to how many seconds?
      a. 10^3 s
      b. 10^6 s
      c. 10^{-3} s
      d. 10^{-6} s
      e.   1 s

ANSWER: c. 10^{-3} s

20.   The fundamental unit of length in the metric system is the
      a.   kilometer
      b.   meter
      c.   centimeter
      d.   gram
      e.   milliliter

ANSWER: b.   meter

21.   The number of milligrams in 1 kg is
      a. 10^3 mg
      b. 10^6 mg
      c. 10^{-3} mg
      d. 10^{-5} mg
      e. 10^{-6} mg

ANSWER: b. 10^6 mg

22.   The SI prefix that corresponds to a factor of 10^{-3} is
      a.   kilo
      b.   deci
      c.   centi
      d.   milli
      e.   none of these

ANSWER: d.   milli

23. Which metric prefix is used to designate 1000?
    a. m
    b. M
    c. k
    d. c
    e. d

    ANSWER: c. k

24. The number of milliliters in 0.0367 liter is
    a. 3.67×10^{-5} mL
    b. 36.7 mL
    c. 3.67 mL
    d. 367 mL
    e. 3, 670 mL

    ANSWER: b. 36.7 mL

25. Which of the following is an SI unit for expressing the mass of a block of Au?
    a. m
    b. g
    c. L
    d. pound

    ANSWER: b. g

26. A cubic centimeter (cm^3) is equivalent to what other metric volume unit?
    a. milliliter
    b. liter
    c. deciliter
    d. centimeter
    e. millimeter

    ANSWER: a. milliliter

27. Convert: 287 mm = _____ m.

    ANSWER: 0.287 m

28. The number of cubic centimeters (cm^3) in 43.0 mL is
    a. $0.0430 \ cm^3$
    b. $4.30 \ cm^3$
    c. $43.0 \ cm^3$
    d. none of these

    ANSWER: c. $43.0 \ cm^3$

29. Using the rules of significant figures, calculate the following:
     6.167 + 70 =
     a. 76
     b. 80
     c. 76.167
     d. 77
     e. 76.17

     ANSWER: b. 80

30. Using the rules of significant figures, calculate the following:
     14.8903 − 2.14 =
     a. 12.7503
     b. 12.75
     c. 12.750
     d. 12
     e. 13

     ANSWER: b. 12.75

31. Using the rules of significant figures, calculate the following:
     12.67 + 13.005 =
     a. 25.675
     b. 25
     c. 20
     d. 25.68
     e. 26

     ANSWER: d. 25.68

32. How many significant figures are in the number 6.022×10^{32}?
     a. 27
     b. 23
     c. 3
     d. 4
     e. 1

     ANSWER: d. 4

33. The number 4.89 rounded to two significant figures is
     a. 4.90
     b. 5.0
     c. 5
     d. 4.9
     e. 4.8

     ANSWER: d. 4.9

34. How many significant figures are in the number 1.89×10^3?
     a. 1
     b. 2
     c. 3
     d. 4
     e. 5

     ANSWER: c.   3

35. The number 2.00152 rounded to four significant figures is
     a. 2.002
     b. 2.001
     c. 2.000
     d. 2.152
     e. none of these

     ANSWER: a.   2.002

36. How many significant figures are in the number 60.02×10^5?
     a. 2
     b. 3
     c. 4
     d. 5
     e. none of these

     ANSWER: c.   4

37. The number 14.809 rounded to three significant figures is
     a. 15.0
     b. 14.9
     c. 14.81
     d. 14.809
     e. 14.8

     ANSWER: e.   14.8

38. Round 23,456 to four significant figures.

     ANSWER: 23,460

39. Round 0.0004583 to three significant figures, and express it in scientific notation.

     ANSWER: 4.58×10^{-4}

40. How many significant figures are in the number 34.00500?
     a. 3
     b. 4
     c. 5
     d. 6
     e. 7

     ANSWER: e.   7

41. How many significant figures are in the number 1000.0?
     a. 1
     b. 2
     c. 3
     d. 4
     e. 5

     ANSWER: e. 5

42. How many significant figures are in the measurement 12.3004 g?
     a. 6
     b. 5
     c. 4
     d. 3
     e. 2

     ANSWER: a. 6

43. How many significant figures are in the number 1.20×10^3?
     a. 1
     b. 2
     c. 3
     d. 4
     e. 5

     ANSWER: c. 3

44. In the sum of 54.34 + 45.66, the number of significant figures is
     a. 2
     b. 3
     c. 4
     d. 5
     e. 6

     ANSWER: d. 5

45. How many significant figures are in the number 3.400?
     a. 1
     b. 2
     c. 3
     d. 4
     e. 5

     ANSWER: d. 4

46. The number 243.306 rounded off to five significant figures is
     a. 243.31
     b. 243.36
     c. 243.30
     d. 243.00
     e. none of these

     ANSWER: a. 243.31

47. A student finds that the weight of an empty beaker is 12.024 g. She places a solid in the beaker to give a combined mass of 12.108 g. To how many significant figures is the mass of the solid known?
    a. 1
    b. 2
    c. 3
    d. 4
    e. 5

    ANSWER: b.   2

48. What is the result of the following multiplication expressed in scientific notation to the correct number of significant figures?
    $(2.08 \times 10^5)(7.0 \times 10^{-5}) =$
    a. 1.4×10^{-1}
    b. 1.456×10^1
    c. 1.5×10^1
    d. 1×10^1
    e. 1.46×10^1

    ANSWER: c. 1.5×10^1

49. How many significant figures are in the number 19.8030?
    a. 6
    b. 5
    c. 4
    d. 3
    e. 2

    ANSWER: a.   6

50. How many significant figures are there in the result of the following calculation?
    $(4.321/2.8) \times (6.9234 \times 10^5)$
    a. 1
    b. 2
    c. 3
    d. 4
    e. 5

    ANSWER: b.   2

51. The result of the following calculation has how many significant figures?
    $(0.4333 \text{ J/g °C}) (33.12°C - 31.12°C)(412.1 \text{ g})$
    a. 1
    b. 2
    c. 3
    d. 4
    e. 5

    ANSWER: c.   3

52. How many significant figures are in the number 4.00700×10^{13}?
    a. 2
    b. 4
    c. 5
    d. 6
    e. none of these

    ANSWER: d. 6

53. How many significant figures are in the number 0.02020×10^{15}?
    a. 3
    b. 4
    c. 5
    d. 6
    e. 19

    ANSWER: b. 4

54. How many significant figures are in the measurement 0.2010 g?
    a. 1
    b. 2
    c. 3
    d. 4
    e. 5

    ANSWER: d. 4

55. The product of $0.1400 \times 6.02 \times 10^{23}$ will have how many significant figures?
    a. 2
    b. 3
    c. 23
    d. 10^{23}
    e. 7

    ANSWER: b. 3

56. How many significant figures should there be in the answer when you divide 4.1 by 7.464?
    a. 7
    b. 4
    c. 3
    d. 2
    e. 1

    ANSWER: d. 2

57. How many significant figures are in the number 0.00204?
    a. 3
    b. 5
    c. 2
    d. 6
    e. 4

    ANSWER: a. 3

58. How many significant figures are in the number 123.00015?
     a. 5
     b. 6
     c. 7
     d. 8
     e. 9

     ANSWER: d. 8

59. How many significant figures are in the number 0.0040090?
     a. 8
     b. 7
     c. 6
     d. 5
     e. 4

     ANSWER: d. 5

60. How many significant figures are in the number 10.050?
     a. 1
     b. 2
     c. 3
     d. 4
     e. 5

     ANSWER: e. 5

61. Write the number 345.607 in scientific notation.

     ANSWER: 3.45607×10^2

62. Write the number 0.0005060 in scientific notation.

     ANSWER: 5.060×10^{-4}

63. Convert 692 m to decimeters.
     a. 69200 dm
     b. 6.92 dm
     c. 69.2 dm
     d. 6920 dm
     e. none of these

     ANSWER: d. 6920 dm

64. Convert 258 L to milliliters.
     a. 0.258 mL
     b. 2.58 mL
     c. 258 mL
     d. 2.58×10^3 mL
     e. 2.58×10^{5} mL

     ANSWER: e. 2.58×10^5 mL

65. Convert 623 qts. to milliliters (1 L - 1.060 qt).
    a. 6.60×10^5 mL
    b. 5.88×10^5 mL
    c. 6.60×10^{-1} mL
    d. 5.88×10^{-1} mL
    e. none of these

    ANSWER: b. 5.88×10^{-5} mL

66. Convert 9.83 kg to pounds (1 lb = 453.6 g).
    a. 21.7 lb
    b. 2.17×10^{-2} lb
    c. 4.46 lb
    d. 4.46×10^6 lb
    e. 4.46×10^3 lb

    ANSWER: a.   21.7 lb

67. Convert 561097 mm to kilometers.
    a. 5.61097 km
    b. 0.561097 km
    c. 561.097 km
    d. 5610.97 km
    e. 5.61097×10^{11} km

    ANSWER: b.   0.561097 km

68. Convert 0.7891 L to centiliters.
    a. 0.007891 cL
    b. 789.1 cL
    c. 78.91 cL
    d. 0.07891 cL
    e. 7.891 cL

    ANSWER: c.   78.91 cL

69. Convert 761 mi to kilometers (1 m = 1.094 yd; 1 mi = 1760 yd).
    a. 4.70×10^{-4} km
    b. 1470 km
    c. 832 km
    d. 696 km
    e. 1220 km

    ANSWER: e.   1220 km

70. Convert 17.6 cm to inches (2.54 cm = 1 in).
    a. 44.7 in
    b. 4.47 in
    c. 6.93 in
    d. 69.3 in
    e. 0.693 in

    ANSWER: c.   6.93 in

71. Convert 0.092 ft^3 to liters (28.32 L = 1 ft^3).
    a. 26.1 L
    b. 2.61 L
    c. 3.2×10^{-3} L
    d. 3.2×10^{-5} L
    e. none of these

    ANSWER: e.   none of these   *B*   *4.3*

72. Convert 6.0 kg to pounds (1 kg = 2.205 lb).
    a. 13 lb
    b. 1.3 lb
    c. 2.7 lb
    d. 27 lb
    e. 0.27 lb

    ANSWER: a.   13 lb

73. Convert: 0.120 mm = _____ m.

    ANSWER: 1.20×10^{-4} m

74. Convert: 0.00110 cm = _____ mm.

    ANSWER: 1.10×10^{-2} mm

75. Convert: 2.00 qt = _____ mL.

    ANSWER: 1890 milliliters

76. Convert: 35.0 cc = _____ mL.

    ANSWER: 35.0 mL

77. Convert: 9.16 qt = _____ mL.

    ANSWER: 8640 mL

78. Convert: 921 qt = _____ mL.

    ANSWER: 8.69×10^5 mL

79.   Convert: 1.83 in = _____ mm.

ANSWER: 46.5 mm

80.   Convert: 42.3 mg = _____ lb.

ANSWER: 9.33×10^{-5} lb

81.   If 1.000 kg equals 2.205 lb, what is the mass in pounds of a human who weighs 75.0 kg?
a.   165.4 lb
b.   34.10 lb
c.   145.9 lb
d.   75.05 lb
e.   none of these

ANSWER: a.   165.4 lb

82.   Convert 62°F to kelvins.
a.   290 K
b.   340 K
c.   303 K
d.   –260 K
e.   –243 K

ANSWER: a.   290 K

83.   Convert: 23.2°C to _____ °F.

ANSWER: 73.8°F

84.   Convert: 801 K = _____ °C.

ANSWER: 528°C

85.   Convert: 16.7°F = _____ °C.

ANSWER: –8.50°C

86.   423 K equals
a.   150.°F
b.   273.°F
c.   696.°F
d.   150.°C
e.   696.°C

ANSWER: d.   150.°C

87.   What Kelvin temperature reading equals 98.6°F?

ANSWER: 310. K

88. Convert: –27.0°F = _____°C.

     ANSWER: –32.8°C

89. Convert: –40.0°C = _____°F.

     ANSWER: –40.0°F

90. 275°F is equivalent to
     a.   230°C
     b.   135°C
     c.   1182.5°C
     d.   120.5°C
     e.   none of these

     ANSWER: b.   135°C

91. Convert: 20.°F = _____°C.

     ANSWER: –6.7°C

92. What is the Celsius equivalent of 481 K?
     a.   208°C
     b.   244°C
     c.   449°C
     d.   718°C
     e.   754°C

     ANSWER: a.   208°C

93. Convert: –27.5°C = _____°F.

     ANSWER: –17.5°F

94. Cesium melts at 302 K and boils at 944 K. What would be the physical state of cesium at 25°C?

     ANSWER: solid

95. Convert: 23°C = _____ K.

     ANSWER: 296 K

96. The Celsius equivalent of 0 K is
     a.   100°C
     b.   0°C
     c.   273°C
     d.   –273°C
     e.   32°C

     ANSWER: d.   –273°C

97. 37°C is approximately equal to
    a. 99°F
    b. 236 K
    c. 37°F
    d. 20°F
    e. none of these

    ANSWER: a.   99°F

98. Convert: 23.9°C = _____°F.

    ANSWER: 75.0°F

99. Convert: 27°C = _____ K.

    ANSWER: 300. K

100. Density is an example of a
     a. chemical property
     b. physical property
     c. qualitative property
     d. chemical change
     e. physical change

     ANSWER: b.   physical property

101. Water has a density of 1.0 g/mL. Which of the objects will float in water?
     Object I: mass - 50.0 g; volume = 60.8 mL
     Object II: mass = 65.2 g; volume = 42.1 mL
     Object III: mass = 100.0 g; volume = 20.0 mL
     a. I
     b. I, III
     c. II
     d. II, III
     e. III

     ANSWER: a.   I

102. Calculate the mass of a rectangular solid that has a density of 2.53 g/cm^{-3} and which measures 2.50 cm by 1.80 cm by 3.00 cm.

     ANSWER: 34.2 g

103. Find the volume of an object that has a density of 3.14 g/ml and a mass of 93.5 g.

     ANSWER: 29.8 mL

104. An experiment requires 75.0 g of ethyl alcohol (density = 0.790 g/mL). What volume, in liters, will be required?

     ANSWER: 9.49 x 10^{-2} L

105. At 20°C the density of mercury is 13.6 g/cm^3. What is the mass of 50.0 mL of mercury at 20°C?

    a. 6.80×10^2 g

    b.   3.68 g

    c.   1.00 g/ml

    d.   0.272 g

    e.   none of these

ANSWER: a. 6.80×10^2 g

106. If a 100.-g sample of platinum metal has a volume of 4.67 mL, what is the density of platinum in g/cm^3?

    a.   21.4 g/cm^3

    b.   2.14 g/cm^3

    c.   0.0467 g/cm^3

    d.   467 g/cm^3

    e.   none of these

ANSWER: a.   21.4 g/cm^3

107. An experiment requires 75.0 mL of ethyl alcohol. If the density of ethyl alcohol is 0.790 g/cm^3, what is the mass of 75.0 mL of ethyl alcohol?

ANSWER: 59.3 g

108. If a 100.-g sample of a metal has a volume of 5.18 mL, what is the density of the metal?

    a.   19.3 g/mL

    b.   1.93 g/mL

    c.   0.518 g/mL

    d.   5 g/mL

    e.   none of these

ANSWER: a.   19.3 g/mL

109. The volume (in milliliters) occupied by 20.0 g of mercury (density = 13.6 g/mL) is

    a.   20.0 x 13.6

    b.   20.0/13.6

    c.   13.6/20.0

    d.   20.0 – 13.6

    e.   none of these

ANSWER: b.   20.0/13.6

110. The density of copper is 8.92 g/mL. The mass of a piece of copper that has a volume of 10.0 mL is
    a. 0.892 g
    b. 892 g
    c. 89.2 g
    d. 8.9×10^2 g
    e. none of these

    ANSWER: c.   89.2 g

111. The density of gold is 19.3 g/mL. What is the volume of a gold nugget that weighs 63.3 g?

    ANSWER: 3.28 mL

112. Aluminum has a density of 2.70 g/cm^3. What is the mass of a rectangular block of aluminum measuring 11.1 cm by 22.2 cm by 33.3 cm?

    ANSWER: 22.2 kg

113. An object has a mass of 40.1 g and occupies a volume of 8.20 mL. The density of this object is
    a. 329 g/mL
    b. 0.204 g/mL
    c. 4.89 g/mL
    d. too low to measure
    e. 40.1 g/mL

    ANSWER: c.   4.89 g/mL

114. What volume would be occupied by a piece of aluminum (density 2.70 g/mL) weighing 85 g?
    a. 229.5 mL
    b. 3.2×10^{-2} mL
    c. 31 mL
    d. 3.2 mL
    e. none of these

    ANSWER: c.   31 mL

115. A graduated cylinder contains 20.0 mL of water. An irregularly shaped object is placed in the cylinder and the water level rises to the 31.2-mL mark. If the object has a mass of 47.9 g, what is its density?

    ANSWER: 4.28 g/mL

116. A piece of an unknown metal weighs 355.2 g and occupies a volume of 72.2 mL. What is the density of this metal?

    ANSWER: 4.92 g/mL

117. A sample of an unknown metal (density = 4.92 g/mL) weighs 922.0 g. What is the volume of this piece of metal?

    ANSWER: 187 mL

118. The density of an object that has a mass of 2.50 g and occupies a volume of 1.20 mL equals
    a. 2.50 g/mL
    b. 1.20 g/mL
    c. 2.08 g/mL
    d. .480 g/mL
    e. 3.00 g/mL

    ANSWER: c.   2.08 g/mL

119. An empty graduated cylinder weighs 55.26 g. When filled with 50.0 mL of an unknown liquid, it weighs 92.39 g. The density of the unknown liquid is
    a. 37.11 g/mL
    b. 50.0 g/mL
    c. 0.743 g/mL
    d. 1.67 g/mL
    e. 0.592 g/mL

    ANSWER: c.   0.743 g/mL

120. A solid object with a volume of 5.62 mL weighs 108 g. Would this object float or sink in mercury? Explain. (Density of Hg = 13.6 g/mL.)

    ANSWER: The object would sink. Density of the object = 19.2 g/mL. *d* (object) > *d*(Hg).

121. Copper has a density of 8.96 g/cm^3. If a cylinder of copper weighing 42.38 g is dropped into a graduated cylinder containing 20.00 mL of water, what will be the new water level?

    ANSWER: 24.73 mL

122. A chemist needs 25.0 g of bromine for an experiment. What volume should she use? (Density of bromine = 3.12 g/cm^3.)

    ANSWER: 8.01 mL

123. A chunk of sulfur has a volume of 7.73 cm^3. What is the mass of this sulfur? (Density of sulfur = 2.07 g/cm^3.)

    ANSWER: 16.0 g

CHAPTER 3

Matter and Energy

1. The state of matter for an object that has neither definite shape nor definite volume is
   a. solid
   b. liquid
   c. gaseous
   d. elemental
   e. mixed

   ANSWER: c. gaseous

2. The state of matter for an object that has a definite volume but not a definite shape is
   a. solid
   b. liquid
   c. gaseous
   d. elemental
   e. mixed

   ANSWER: b. liquid

3. The state of matter for an object that has both definite volume and definite shape is
   a. solid
   b. liquid
   c. gaseous
   d. elemental
   e. mixed

   ANSWER:·   a. solid

4. Anything that has mass and volume is called _____.

   ANSWER: matter

5. Classify each of the following as a physical (P) or a chemical (C) change.

   _____ a.   cooking an egg
   _____ b.   boiling water
   _____ c.   ironing a shirt
   _____ d.   burning gasoline
   _____ e.   decomposing water
   _____ f.   evaporating alcohol
   _____ g.   sanding a table top
   _____ h.   grinding grain
   _____ i.   fermenting fruit juice
   _____ j.   dissolving sugar in water

   ANSWER:
   a. C
   b. P
   c. P
   d. C
   e. C
   f. P
   g. P
   h. P
   i. C
   j. P

6. Which of the following involves a chemical change?
   a. boiling water
   b. melting ice
   c. chopping wood
   d. cooking an egg
   e. none of these

   ANSWER: d. cooking an egg

7. Which of the following is a physical change?
   a. burning gasoline
   b. cooking an egg
   c. decomposing meat
   d. evaporating water
   e. rusting iron

   ANSWER: d. evaporating water

8. Which of these is a chemical property?
   a. Ice melts at 0°C.
   b. Oxygen is a gas.
   c. Helium is very nonreactive.
   d. Sodium is a soft, shiny metal.
   e. Water has a high specific heat.

   ANSWER: c. Helium is very nonreactive.

9. Which of the following involves no chemical change?
    a. burning paper
    b. boiling water
    c. baking a cake
    d. lighting a match
    e. driving a car

    ANSWER: b.   boiling water

10. Which of the following is only a physical change?
    a. Sugar dissolves in coffee.
    b. Cookies burn in the oven.
    c. A banana ripens.
    d. Leaves turn colors in the fall.
    e. At least two of the above (a-d) exhibit only a physical change.

    ANSWER: a.   Sugar dissolves in coffee.

11. Which of the following is a chemical change?
    a. Water condenses on a mirror.
    b. A damp towel dries.
    c. Peanuts are crushed.
    d. A "tin" can rusts.
    e. At least two of the above (a-d) exhibit a chemical change.

    ANSWER: d.   A "tin" can rusts.

12. An example of a chemical change is
    a. boiling alcohol
    b. grinding coffee beans.
    c. digesting a pizza
    d. coffee spilled on a shirt
    e. an ice cube melting in a drink

    ANSWER: c.   digesting a pizza

13. In a chemical change,
    a. a phase change must occur
    b. the original material can never be regenerated
    c. a phase change never occurs
    d. the products are different substances from the starting materials

    ANSWER: d.   the products are different substances from the starting materials

14. All _____ changes are _____ changes.

    ANSWER: chemical; physical

15. If iodine melts at 114°C and boils at 184°C, what is its physical state at 120°C?

    ANSWER: liquid

16. If iodine melts at 114°C and boils at 184°C, what is its physical state at 98°C?

     ANSWER: solid

17. If iodine melts at 114°C and boils at 184°C, what is its physical state at 250°C?

     ANSWER: gas

18. Which of the following describes a chemical property of gold?
     a. Gold is a yellow metal.
     b. Gold is an inert (nonreactive) metal.
     c. Gold is a soft metal.
     d. Gold is a very dense metal.
     e. Gold is a good conductor of heat and electricity.

     ANSWER: b.   Gold is an inert (nonreactive) metal.

19. Which of the following is a chemical change?
     a. water boiling
     b. gasoline evaporating
     c. butter melting
     d. sugar dissolving in water
     e. paper burning

     ANSWER: e.   paper burning

20. How many of the following are pure compounds: sodium, sugar, oxygen, air, iron?
     a. 1
     b. 2
     c. 3
     d. 4
     e. 5

     ANSWER: a.   1

21. A(n) _____ always has the same composition.

     ANSWER: compound

22. Which of the following is an element?
     a. air
     b. water
     c. salt
     d. helium
     e. sugar

     ANSWER: d.   helium

23. True or false? A compound can consist of one kind of element.

     ANSWER: False

24. Which of the following is an element?
    a. brass
    b. salt
    c. water
    d. earth
    e. oxygen

    ANSWER: e. oxygen

25. Which of these is an element?
    a. water
    b. iron ore
    c. wood
    d. silver
    e. brass

    ANSWER: d. silver

26. An example of a mixture is
    a. hydrogen fluoride
    b. purified water
    c. gold
    d. the air in this room
    e. all of these

    ANSWER: d. the air in this room

27. An example of a pure substance is
    a. elements
    b. compounds
    c. pure water
    d. carbon dioxide
    e. all of these

    ANSWER: e. all of these

28. Which of the following is an incorrect description?
    a. A homogeneous mixture.
    b. A heterogeneous compound.
    c. A solid element.
    d. A mixture of solids.
    e. A solution of gases.

    ANSWER: b. A heterogeneous compound.

29. A homogeneous mixture is also called _____.
    a. a heterogeneous mixture.
    b. a pure substance.
    c. a compound.
    d. a solution.
    e. an element.

    ANSWER: d. a solution

30. Classify each of the following as an element (A), a compound (B), a homogeneous mixture (C), or a heterogeneous mixture (D).
    a. table salt   _____
    b. chlorine gas   _____
    c. sand in water   _____
    d. petroleum   _____
    e. caffeine   _____

    ANSWER:
    a. C
    b. A
    c. D
    d. C
    e. B

31. Water is an example of
    a. a homogeneous mixture
    b. a heterogeneous mixture
    c. a compound
    d. an element

    ANSWER: c.   a compound

32. A solution can be distinguished from a compound by its
    a. variable composition
    b. liquid state
    c. heterogeneous nature
    d. lack of color

    ANSWER: a.   variable composition

33. Which of the following is an element?
    a. iron
    b. wood
    c. water
    d. blood
    e. none of these

    ANSWER: a.   iron

34. Classify each of the following as an element (E), a compound (C), or a mixture (M).
    a. 14K gold   _____
    b. pure silver   _____
    c. aluminum   _____
    d. distilled water   _____
    e. tap water   _____
    f. brass   _____
    g. tungsten   _____
    h. sodium chloride   _____
    i. air   _____

    ANSWER:
    a. M
    b. E
    c. E
    d. C
    e. M
    f. M
    g. E
    h. C
    i. M

35. Which would be an example of a homogeneous mixture?
    a. vodka
    b. oily water
    c. soil (dust)
    d. sodium chloride
    e. aluminum

    ANSWER: a. vodka

36. Helium is an example of
    a. a homogeneous mixture
    b. a heterogeneous mixture
    c. a compound
    d. an element

    ANSWER: d. an element

37. The process of filtering a sand-saltwater mixture is a _____ process.

    ANSWER: physical

38. Which of the following processes require(s) chemical methods?
    a. Separating a homogeneous mixture into pure substances.
    b. Separating a heterogeneous mixture into pure substances.
    c. Distilling a saltwater mixture.
    d. Breaking a compound into its constituent elements.
    e. At least two of the above (a-d) require chemical methods.

    ANSWER: d. Breaking a compound into its constituent elements.

39. The amount of energy needed to heat 2.00 g of carbon from 50.0°C to 80.0°C is 42.6 J. The specific heat capacity of this sample of carbon is
    a.   2556 J/g °C
    b.   0.710 J/g °C
    c.   639 J/g °C
    d.   0.355 J/g °C
    e.   1.42 J/g °C

    ANSWER: b.   0.710 J/g °C

40. A 5.10-g sample of iron is heated from 36.0°C to 75.0°C. The amount of energy required is 89.5 J. The specific heat capacity of this sample of iron is
    a.   17800 J/g °C
    b.   0.900 J/g °C
    c.   11.7 J/g °C
    d.   0.450 J/g °C
    e.   2.22 J/g °C

    ANSWER: d.   0.450 J/g °C

41. A 6.75-g sample of gold (specific heat capacity = 0.130 J/g °C) is heated using 50.6 J of energy. If the original temperature of the gold is 25.0°C, what is its final temperature?
    a.   82.7°C
    b.   57.7°C
    c.   24.4°C
    d.   43.4°C
    e.   32.7°C

    ANSWER: a.   82.7°C

42. Which of the following is a valid unit for specific heat (or specific heat capacity)?
    a.   cal/g °C
    b.   cal
    c.   cal/g
    d.   °C
    e.   g °C/cal

    ANSWER: a.   cal/g °C

43. Heat is typically measured in
    a.   °C
    b.   °F
    c.   joules
    d.   grams

    ANSWER: c.   joules

44. Calculate the heat given off when 177 g of copper cools from 155.0°C to 23.0°C. The specific heat capacity of copper is 0.385 J/g °C.

    ANSWER: 9.00 x 10^4 J

45. SI units for specific heat capacity are
    a. cal
    b. cal/g
    c. J/g °C
    d. g/mL
    e. g °C/cal

    ANSWER: c.   J/g °C

46. The quantity of heat required to change the temperature of 1 g of a substance by 1°C is defined as
    a. a joule
    b. specific heat capacity
    c. a calorie
    d. density

    ANSWER: b.   specific heat capacity

47. The specific heat capacity of iron is 0.45 J/g °C. How many joules of energy are needed to warm 1.50 g of iron from 20.00°C to 29.00°C?

    ANSWER: 6.1 J

48. How many joules of energy would be required to heat 24.2 g of carbon from 23.6°C to 54.2°C? (Specific heat capacity of carbon = 0.71 J/g °C.)

    ANSWER: 5.3×10^2 J

49. The specific heat capacity of silver is 0.24 J/g °C. How many joules of energy are needed to warm 0.500 g of silver from 25.0°C to 27.5°C?

    ANSWER: 0.30 J

50. The specific heat capacity of aluminum is 0.89 J/g °C. Calculate the amount of energy needed to warm 1.92×10^3 g of aluminum from 73.0°C to 155.0°C?

    ANSWER: 140 kJ

51. Assume that 372 J of heat is added to 5.00 g of water originally at 23.0°C. What would be the final temperature of the water? (Specific heat capacity of water = 4.184 J/g °C.)

    ANSWER: 40.8°C

52. How much energy will be needed to heat 60.0 gal of water from 22.0°C to 110.0°C? (Note that 1.00 gal weighs 3.77 kg and that water has a specific heat capacity of 4.184 J/g °C.)

    ANSWER: 8.33×10^4 kJ

53. Which is larger, one calorie or one joule?

    ANSWER: one calorie

CHAPTER 4

Chemical Foundations: Elements, Atoms, and Ions

1. What is the most abundant element on the earth (including the crust, oceans, and atmosphere)?
   a. silicon
   b. oxygen
   c. hydrogen
   d. carbon
   e. iron

   ANSWER: b.   oxygen

2. What is the most abundant element found in the human body?
   a. carbon
   b. hydrogen
   c. calcium
   d. oxygen
   e. water

   ANSWER: d.   oxygen

3. The symbol for the element chlorine is
   a. C
   b. Ch
   c. C o
   d. Cs
   e. Cl

   ANSWER: e.   Cl

4. The symbol for the element potassium is
   a. Po
   b. P
   c. Pt
   d. K
   e. Pm

   ANSWER: d.   K

5. The symbol for the element cobalt is
    a. C
    b. Cu
    c. Co
    d. Cb
    e. K

    ANSWER: c. Co

6. The symbol for the element mercury is
    a. Hg
    b. Mn
    c. Mg
    d. Ag
    e. Mr

    ANSWER: a. Hg

7. The symbol for the element lithium is
    a. L
    b. Li
    c. Lt
    d. Lm
    e. Lh

    ANSWER: b. Li

8. The symbol C stands for the element
    a. cobalt
    b. copper
    c. chromium
    d. carbon
    e. Two of the above are correct.

    ANSWER: d. carbon

9. The symbol Ag can be used to represent
    a. a silver ion
    b. a silver atom
    c. an aluminum ion
    d. an aluminum atom
    e. gold

    ANSWER: b. a silver atom

10. The symbol for beryllium is
    a. Be
    b. B
    c. Br
    d. Ber
    e. Bl

    ANSWER: a. Be

11. Give the symbols for the following elements.
    a. copper _____
    b. sodium _____
    c. chlorine _____

    ANSWER:
    a. Cu
    b. Na
    c. Cl

12. Give the names that correspond to the following.
    a. Au _____
    b. S _____
    c. Ba _____

    ANSWER:
    a. gold
    b. sulfur
    c. barium

13. S is the symbol for
    a. sodium
    b. selenium
    c. sulfur
    d. silicon

    ANSWER: c.   sulfur

14. The symbol for manganese is
    a. Mg
    b. Mn
    c. Ma
    d. Mo
    e. none of these

    ANSWER: b.   Mn

15. Four of the most important elements on this planet are hydrogen, oxygen, carbon, and nitrogen. Give the symbol for each of these elements.
    a. hydrogen _____
    b. oxygen _____
    c. carbon _____
    d. nitrogen _____

    ANSWER:
    a. H
    b. O
    c. C
    d. N

16. The symbol for magnesium is
    a. Mg
    b. Mn
    c. Ma
    d. Mo
    e. none of these

    ANSWER: a.   Mg

17. Write the name for Cs. _____

    ANSWER cesium

18. Write the name for K. _____

    ANSWER: potassium

19. Write the name for Kr. _____

    ANSWER: krypton

20. Write the name for Zr. _____

    ANSWER: zirconium

21. Write the name for Ag. _____

    ANSWER:silver

22. Write the symbol for cobalt. _____

    ANSWER: Co

23. Write the symbol calcium. _____

    ANSWER: Ca

24. Write the symbol for carbon. _____

    ANSWER: C

25. Write the symbol for manganese. _____

    ANSWER: Mn

26. Write the symbol for lead. _____

    ANSWER: Pb

27. The symbol for the element silver is
    a. Si
    b. S
    c. Au
    d. Ar
    e. Ag

    ANSWER: e. Ag

28. Co is the symbol for
    a. carbon
    b. cobalt
    c. carbon monoxide
    d. carboxide
    e. chlorine

    ANSWER: b. cobalt

29. The symbol for nitrogen is
    a. Na
    b. N
    c. Ng
    d. Ni
    e. none of these

    ANSWER: b. N

30. The symbol for copper is
    a. Cr
    b. C
    c. Co
    d. Cu
    e. none of these

    ANSWER: d. Cu

31. Si is the symbol for
    a. silver
    b. silicon
    c. selenium
    d. sodium
    e. sulfur

    ANSWER: b. silicon

32. How many of the following did Dalton not discuss in his atomic theory?
    isotopes
    ions
    protons
    electrons
    neutrons
    a. 1
    b. 2
    c. 3
    d. 4
    e. 5

    ANSWER: e. 5

33. The law of constant composition applies to
    a. metals
    b. metalloids
    c. homogeneous mixtures
    d. heterogeneous mixtures
    e. compounds

    ANSWER: e. compounds

34. A substance composed of two or more elements combined chemically in a fixed proportion by mass is
    a. a compound
    b. a mixture
    c. an atom
    d. a solid
    e. none of these

    ANSWER: a. a compound

35. The fundamental "particle" of a chemical element according to Dalton's theory is the
    a. electron
    b. molecule
    c. atom
    d. compound

    ANSWER: c. atom

36. In chemistry, a formula is used to represent
    a. a compound
    b. an element
    c. a metal
    d. all of these
    e. none of these

    ANSWER: a. a compound

37. How many hydrogen atoms are indicated in the formula $(NH_4)_2C_8H_4O_2$?

    a. 8
    b. 12
    c. 20
    d. 24
    e. none of these

    ANSWER: b. 12

38. The total number of oxygen atoms indicated by the formula $Fe_2(CO_3)_3$ is

    a. 3
    b. 6
    c. 9
    d. 12
    e. 18

    ANSWER: c. 9

39. How many atoms of hydrogen are in one molecule of CH_3Cl?

    a. 6
    b. 6×10^{23}
    c. 3
    d. 30×10^{23}
    e. 18×10^{23}

    ANSWER: c. 3

40. How many atoms are represented by one formula unit of aluminum dicromate, $Al_2(Cr_2O_7)_3$?

    a. 14
    b. 25
    c. 27
    d. 29
    e. none of these

    ANSWER: d. 29

41. How many phosphorus atoms are represented by one formula unit of calcium phosphate, $Ca_3(PO_4)_3$?

    a. 3
    b. 6
    c. 9
    d. 12
    e. 18

    ANSWER: a. 3

42. How many atoms of oxygen are in one formula unit of calcium hydrogen sulfate?
    a. 3
    b. 4
    c. 5
    d. 6
    e. 8

    ANSWER: b. 4

43. The total number of atoms indicated by the formula $Ca_3(PO_4)_3$ is
    a. 6
    b. 10
    c. 16
    d. 18
    e. 7

    ANSWER: d. 18

44. How many nitrogen atoms are indicated by the formula $Al(NO_3)_3$?
    a. 1
    b. 3
    c. 9
    d. 4
    e. 0

    ANSWER: b. 3

45. Mendeleev is credited with the modern atomic theory of atoms. T/F _____

    ANSWER: False

46. Which of the following contains the largest number of oxygen atoms?
    a. $4H_2O$
    b. $3CO_2$
    c. O_3
    d. H_2SO_4
    e. $Al(NO_3)_3$

    ANSWER: e. $Al(NO_3)_3$

47. The chemical formula Al_2O_3 indicates
    a. two atoms of aluminum and three atoms of oxygen
    b. three atoms of aluminum and two atoms of oxygen
    c. six atoms of each element
    d. five atoms of each element
    e. None of these is correct.

    ANSWER: a. two atoms of aluminum and three atoms of oxygen

48. The chemical formula Ga_2O_3 indicates
     a. two atoms of gallium and three atoms of oxygen
     b. three atoms of gallium and two atoms of oxygen
     c. six atoms of each element
     d. five atoms of each element
     e. none of these

     ANSWER: a. two atoms of gallium and three atoms of oxygen

49. The first scientist to show that atoms emit tiny negative particles was
     a. J. J. Thomson
     b. Lord Kelvin
     c. Ernest Rutherford
     d. William Thomson
     e. James Chadwick

     ANSWER: a. J. J. Thomson

50. The scientist whose alpha-particle scattering experiment led him to conclude that the nucleus of an atom contains a dense center of positive charge is
     a. J. J. Thomson
     b. Lord Kelvin
     c. Ernest Rutherford
     d. William Thomson
     e. James Chadwick

     ANSWER: c. Ernest Rutherford

51. Which atomic particle determines the chemical behavior of an atom?
     a. proton
     b. electron
     c. neutron
     d. nucleus
     e. none of these

     ANSWER: b. electron

52. List the three main subatomic particles.

     ANSWER: electron, proton, neutron

53. Which particle has the smallest mass?
     a. neutron
     b. proton
     c. electron
     d. helium nucleus

     ANSWER: c. electron

54. How many protons, electrons, and neutrons respectively does ^{127} I have?

    a. 53, 127, 74
    b. 53, 74, 53
    c. 53, 53, 127
    d. 74, 53, 127
    e. 53, 53, 74

    ANSWER:　　e.　53, 53, 74

55. Which of the following statements are true?
    I.　The number of protons in an element is the same for all neutral atoms of that element.
    II.　The number of electrons in an element is the same for all neutral atoms of that element.
    III.　The number of neutrons in an element is the same for all neutral atoms of that element.
    a.　I, II and III are all true.
    b.　Only I and II are true.
    c.　Only II and III are true.
    d.　Only I and II are true.
    e.　I, II and III are all false.

    ANSWER:　　b.　Only I and II are true.

56. An element's most stable ion forms an ionic compound with chlorine having the formula XCl_2.
    If the ion of element X has a mass of 89 and 36 electrons, what is the identity of the element, and how many neutrons does it have?
    a.　Kr, 53 neutrons
    b.　Kr, 55 neutrons
    c.　Se, 55 neutrons
    d.　Sr, 51 neutrons
    e.　Rb, 52 neutrons

    ANSWER:　　d.　Sr, 51 neutrons

57. How many protons, electrons, and neutrons, respectively, does ^{16} O have?

    a.　8, 18, 8
    b.　8, 8, 8
    c.　8, 10, 8
    d.　8, 14, 8
    e.　8, 18, 16

    ANSWER:　　b.　8, 8, 8

58. The number of neutrons in one atom of $^{206}_{82}$ Hg is

    a.　82
    b.　206
    c.　124
    d.　288
    e.　none of these

    ANSWER:　　c.　124

59. An atom with 15 protons and 16 neutrons is an atom of
     a. P
     b. Ga
     c. S
     d. Pd
     e. Rh

     ANSWER: a. P

60. How many protons, electrons, and neutrons, respectively, does $^{27}Al^{3+}$ have?

     a. 13, 13, 14
     b. 13, 10, 14
     c. 13, 13, 27
     d. 13, 10, 27
     e. 13, 13, 13

     ANSWER: b. 13, 10, 14

61. Identify each of the following.
     a. The heaviest noble gas
     b. The transition metal that has 24 electrons as a +3 ion
     c. The halogen in the third period
     d. The alkaline earth metal that has 18 electrons as a stable ion

     ANSWER:
     a. Rn
     b. Co
     c. Cl
     d. Ca

62. The average mass of a boron atom is 10.81. Assuming you were able to isolate only one boron atom, the chance that you would randomly get one with a mass of 10.81 is
     a. 0%
     b. 0.81%
     c. about 11%
     d. 10.81%
     e. greater than 50%

     ANSWER: a. 0%

63. Atoms of the same element having the same atomic number but different mass numbers are called
     a. isomers
     b. orbitals
     c. neutrons
     d. isotopes
     e. nuclei

     ANSWER: b. isotopes

64. The atom with 68 neutrons and 50 protons has a mass number of
    a. 68
    b. 50
    c. 18
    d. 118
    e. cannot be determined from information given

    ANSWER: d. 118

65. What is the element and number of neutrons for an element with atomic number 26 and mass number 56?.

    ANSWER: Fe, 30 neutrons

66. Determine the number of protons and neutrons in Cu-63.

    ANSWER: 29 protons, 34 neutrons

67. How many neutrons are there in one atom of $^{48}_{22}$Ti?

    a. 22
    b. 26
    c. 48
    d. 70
    e. none of these

    ANSWER: b. 26

68. The mass number of an atom equals
    a. the number of neutrons per atom
    b. the atomic mass of the element
    c. the atomic number of the element
    d. the number of protons plus the number of neutrons per atom
    e. none of the above

    ANSWER: d.   the number of protons plus the number of neutrons per atom

69. How many neutrons are contained in an iodine nucleus with a mass number of 131?
    a. 53
    b. 74
    c. 78
    d. 127
    e. 131

    ANSWER: c.   78

42 *Chapter 4*

70. Which of these are isotopes of hydrogen?
    a. ^{12}C and ^{13}C
    b. ^2H and He
    c. ^{42}K and ^{40}K
    d. ^3H and ^2H
    e. Li· and He

    ANSWER: d. ^3H and ^2H

71. How many electrons are present in a tin atom with a mass number of 119?
    a. 50
    b. 119
    c. 169
    d. 69
    e. 6.02×10^{23}

    ANSWER: a.   50

72. The number of protons in $^{200}_{80}Hg$ is

    a. 80
    b. 120
    c. 200
    d. dependent on ionic charge
    e. unknown

    ANSWER: a.   80

73. How many electrons are present in a bromine atom with a mass number of 87?
    a. 35
    b. 87
    c. 122
    d. 80
    e. 6.02×10^{23}

    ANSWER: a.   35

74. An atom that has 54 protons, 54 electrons, and 78 neutrons is

    a. $^{132}_{54}Xe$

    b. $^{132}_{55}Cs$

    c. $^{78}_{54}Xe$

    d. $^{54}_{78}Pt$

    e. $^{108}_{78}Pt$

    ANSWER: a. $^{132}_{54}Xe$

75. The number of protons in the nucleus of an atom is called its
     a. mass number
     b. valence
     c. isotope number
     d. atomic number
     e. none of these

     ANSWER: d. atomic number

76. The number of neutrons in $^{7}_{3}$ Li is

     a. 10
     b. 8
     c. 7
     d. 4
     e. 3

     ANSWER: d. 4

77. Which pair has approximately the same mass?
     a. a hydrogen, $^{1}_{1}$ H, and a deuterium, $^{2}_{1}$ H, atom

     b. a neutron and an electron
     c. a proton and a neutron
     d. an electron and a proton

     ANSWER: c. a proton and a neutron

78. Choose the pair in which the components have the same charge.
     a. a proton and a hydrogen atom
     b. a hydrogen atom and an electron
     c. a neutron and a hydrogen atom
     d. an electron and a neutron

     ANSWER: c. a neutron and a hydrogen atom

79-84. Determine the number of protons and neutrons for each of the following.

79. Rb-85

     ANSWER: 37 protons, 48 neutrons

80. Sn-118

     ANSWER: 50 protons, 68 neutrons

81. C-14

     ANSWER: 6 protons, 8 neutrons

82. I-127

     ANSWER: 53 protons, 74 neutrons

83.   Pb-207

ANSWER: 82 protons, 125 neutrons

84.   Hg-201

ANSWER: 80 protons, 121 neutrons

85. ^{15}N and ^{15}O are isotopes. T/F _____

ANSWER: False

86.   Hydrogen has four naturally occurring isotopes. T/F _____

ANSWER: False

87.   Give the name and symbol for the element that has mass number 64 and 35 neutrons.

ANSWER: copper, $^{64}_{29}$Cu

88.   Classify the following as a metal, nonmetal, or metalloid.
    a.   arsenic _____
    b.   argon _____
    c.   calcium _____
    d.   plutonium _____

ANSWER:
    a.   metalloid
    b.   nonmetal
    c.   metal
    d.   metal

89.   Which of the following is a nonmetal?
    a.   cerium
    b.   cesium
    c.   carbon
    d.   calcium
    e.   copper

ANSWER: c.   carbon

90. For the element sulfur, $^{32}_{16}$S, give the following.

    a. atomic number _____
    b. mass number _____
    c. number of neutrons _____
    d. number of electrons _____
    e. number of protons _____

    ANSWER:
    a. 16
    b. 32
    c. 16
    d. 16
    e. 16

91. How many protons are there in an atom of $^{238}_{92}$U?

    a. 92
    b. 238
    c. 146
    d. 184
    e. Cannot tell from information given

    ANSWER: a.   92

92. An atom with 45 protons has a mass number of 100. It must contain how many neutrons?
    a. 145
    b. 45
    c. 100
    d. 55
    e. none of these

    ANSWER: d.   55

93. An atom with 46 protons has a mass number of 102. The atom is
    a. Ba
    b. Pd
    c. No
    d. U
    e. none of these

    ANSWER: b.   Pd

94. How many electrons are present in a fluorine, F, atom?
    a. 9
    b. 10
    c. 11
    d. 18
    e. 19

    ANSWER: a. 9

95. How many neutrons are present in a $^{234}_{91}$ Pa nucleus?

    a. 91
    b. 143
    c. 234
    d. 325
    e. none of these

    ANSWER: b. 143

96. How many protons are in the nucleus of a Mg atom?
    a. 12
    b. 24
    c. 25
    d. 36
    e. variable

    ANSWER: a. 12

97. How many protons are in the atom $^{14}_{6}$C?

    a. 14
    b. 6
    c. 8
    d. 20
    e. none of these

    ANSWER: b. 6

98. Which of the following elements is an alkali metal?
    a. Ca
    b. Cu
    c. Fe
    d. Na
    e. Sc

    ANSWER: d. Na

99. Which of the following elements is an alkaline earth metal?
    a. Ca
    b. Cu
    c. Fe
    d. Na
    e. Sc

    ANSWER: a. Ca

100. Which of the following is a noble gas?
      a. Ar
      b. N_2
      c. H_2
      d. O_2
      e. CO_2

      ANSWER: a. Ar

101. Which of the following elements is most similar to chlorine?
      a. H
      b. He
      c. Na
      d. Hg
      e. Br

      ANSWER: e. Br

CHAPTER 5a

Nomenclature

1. Compounds containing only two chemical elements are called
   a. mixtures
   b. binary compounds
   c. ternary compounds
   d. heterogeneous substances
   e. none of these

   ANSWER: b. binary compounds

2. Which of the following is a binary compound?
   a. O_2
   b. HCN
   c. H_2SO_4
   d. H_2S
   e. NaOH

   ANSWER: d. H_2S

3. The correct name for LiCl is
   a. lithium monochloride
   b. lithium(I) chloride
   c. monolithium chloride
   d. lithium chloride
   e. monolithium monochloride

   ANSWER: d. lithium chloride

4. The name for Hg_2^{2+} is

   a. mercury(I) ion
   b. mercury ion
   c. mercury(II) ion
   d. hydrogen ion
   e. hydrogen(II) ion

   ANSWER: a. mercury(I) ion

5. The correct name for FeO is
   a. iron oxide
   b. iron(II) oxide
   c. iron(III) oxide
   d. iron monoxide
   e. iron(I) oxide

   ANSWER: b.   iron(II) oxide

6. The charge on a potassium ion in its ionic compounds is
   a. 1+
   b. 2+
   c. Various charges are possible.

   ANSWER: a.   1+

7. The symbol for the calcium ion is
   a. C^{2+}
   b. Ca^{2+}
   c. Cl^{2+}
   d. Ca
   e. Ca^+

   ANSWER: b. Ca^{2+}

8. Na^+ represents a
   a. sodium ion
   b. sodium atom
   c. potassium ion
   d. potassium atom
   e. naval ion

   ANSWER: a.   sodium ion

9. The correct name for the Ca^{2+} species is
   a. calcium
   b. calcium(II) ion
   c. calcium ion
   d. calcium(I) ion
   e. monocalcium ion

   ANSWER: c.   calcium ion

10. Which of the following is *not* the correct chemical formula for the compound named?
    a. hydrocyanic acid   HCN*(aq)*
    b. calcium sulfate $CaSO_4$
    c. beryllium oxide   BeO
    d. nickel(II) peroxide Ni_2O
    e. ammonium chromate $(NH_4)_2CrO_4$

    ANSWER: d.   nickel(II) peroxide

11. Which of the following formulas is *incorrect*?
    a. $Ba(OH)_2$
    b. LiH
    c. CaCl
    d. $KMnO_4$
    e. K_2O

    ANSWER: c. CaCl

12. Titanium(IV) oxide has the formula
    a. Ti_4O
    b. TiO_4
    c. $Ti(IV)O$
    d. TiO_2
    e. Ti_4O_2

    ANSWER: d. TiO_2

13. The name for Li_2O is _____.

    ANSWER: lithium oxide

14. The correct name for P_2O_5 is
    a. phosphorus(II) oxide
    b. phosphorus(V) oxide
    c. diphosphorus oxide
    d. diphosphorus pentoxide
    e. phosphorus pentoxide

    ANSWER: d. diphosphorus pentoxide

15. CO is the formula for
    a. copper
    b. cobalt
    c. carbon monoxide
    d. none of these

    ANSWER: c. carbon monoxide

16. Which is the formula for nitrous oxide?
    a. NO
    b. N_2O
    c. NO_3
    d. N_2O_3
    e. NO_2

    ANSWER: b. N_2O

17. The binary compound PCl_3 is called
    a. phosphorus chloride
    b. triphosphorus chloride
    c. monophosphorus trichloride
    d. phosphorus trichloride
    e. none of these

    ANSWER: d. phosphorus trichloride

18. The compound PI_3 is named
    a. potassium iodide
    b. monophosphorus iodide
    c. phosphorus iodide
    d. phosphorus triiodide
    e. none of these

    ANSWER: d. phosphorus triiodide

19. The name for SO_2 is _____.

    ANSWER: sulfur dioxide

20. The formula for calcium hydrogen sulfate is
    a. $Ca(SO_4)_2$
    b. CaS_2
    c. $Ca(HSO_4)_2$
    d. Ca_2HSO_4
    e. Ca_2S

    ANSWER: c. $Ca(HSO_4)_2$

21. The name of the compound $Pb(NO_3)_2$ is
    a. lead nitrate
    b. lead dinitrate
    c. lead(II) dinitrate
    d. lead(II) nitrite
    e. lead(II) nitrate

    ANSWER: e. lead(II) nitrate

22. The name of the compound NH_4ClO_4 is
    a. ammonium perchlorate
    b. ammonium chlorite
    c. ammonium chlorate
    d. ammonium hypochlorate
    e. ammonium chloride

    ANSWER: a. ammonium perchlorate

23. The correct formula for the carbonate ion is
    a. CO_3^-
    b. CO_3^{2-}
    c. CO_4^-
    d. CO_4^{2-}
    e. CO_3^{3-}

    ANSWER: b. CO_3^{2-}

24. The correct formula for ammonium sulfate is
    a. NH_4SO_3
    b. NH_4SO_4
    c. $(NH_4)_2SO_3$
    d. $(NH_4)_2SO_4$
    e. $(NH_3)_2SO_3$

    ANSWER: d. $(NH_4)_2SO_4$

25. The name for the NO_3^- ion is
    a. nitrate ion
    b. nitrite ion
    c. nitrogen ion
    d. nitric ion
    e. nitrous ion

    ANSWER: a.   nitrate ion

26. The correct name for an aqueous solution of H_2SO_4 is
    a. sulfurous acid
    b. hydrosulfurous acid
    c. sulfuric acid
    d. hydrosulfuric acid
    e. none of these

    ANSWER: c.   sulfuric acid

27. The correct name for an aqueous solution of H_2CO_3 is
    a. carbonate acid
    b. hydrocarbonic acid
    c. carbonous acid
    d. carbonic acid
    e. hydrocarbonous acid

    ANSWER: d.   carbonic acid

28. The correct name for an aqueous solution of H_3PO_4 is
     a. hydrophosphoric acid
     b. phosphorous acid
     c. phosphate acid
     d. hydrophosphorus acid
     e. phosphoric acid

     ANSWER: e.   phosphoric acid

29. The correct name for an aqueous solution of HCN is
     a. hydrocyanic acid
     b. cyanic acid
     c. cyanate acid
     d. hydrocyanous acid
     e. cyanous acid

     ANSWER: a.   hydrocyanic acid

30. The correct name for an aqueous solution of HCl is
     a. chloric acid
     b. hydrochloric acid
     c. hypochloric acid
     d. hypochlorous acid
     e. perchloric acid

     ANSWER: b.   hydrochloric acid

31. The correct name for an aqueous solution of HCN is
     a. hydrogen cyanide
     b. hydrocyanic acid
     c. hydrogen cyanic acid
     d. hydrogen cyanous acid
     e. hydrogen cyanide acid

     ANSWER: b.   hydrocyanic acid

32. The correct name for an aqueous solution of $HC_2H_3O_2$ is
     a. hydrocarbonate acid
     b. hydrocarbonic acid
     c. hydroacetic acid
     d. acetic acid
     e. none of these

     ANSWER: d.   acetic acid

33. The name for ClO_4^- is _____.

     ANSWER: perchlorate ion

34. The correct name for the Sn^{4+} species is _____.

     ANSWER: tin(IV) ion

35. The name for the Fe^{2+} species is _____.

     ANSWER: iron(II) ion

36. The name for $CO_3{}^{2-}$ is _____.

     ANSWER: carbonate ion

37. The name for $NO_3{}^-$ is _____.

     ANSWER: nitrate ion

38. The correct formula for the ammonium ion is
     a. $NH_4{}^{2+}$
     b. N_4H^+
     c. $NH_4{}^+$
     d. NH_4
     e. Am^+

     ANSWER: c. $NH_4{}^+$

39. The carbonate ion has the formula $CO_3{}^{2-}$. What is the correct formula for sodium carbonate?
     a. $Na(CO_3)_2$
     b. $Na_2(CO_3)_2$
     c. Na_2CO_3
     d. $Na_3(CO)_2$
     e. $NaCO_3$

     ANSWER: c. Na_2CO_3

40. What is the name for $HPO_4{}^{2-}$?
     a. phosphate ion
     b. phosphite ion
     c. hydrogen phosphate ion
     d. hydrogen phosphite ion
     e. hydrogen phosphorus oxide ion

     ANSWER: c.   hydrogen phosphate ion

41. The substance $ClO_3{}^-$ is best described as
     a. a molecule
     b. a polyatomic ion
     c. a polyatomic molecule
     d. a mixture

     ANSWER: b.   a polyatomic ion

42. The formula for the compound formed from the polyatomic ions NH_4^+ and PO_4^{3-} is
    a. $NH_4(PO_4)_3$
    b. NH_4PO_4
    c. $(NH_4)_3PO_4$
    d. $(NH_4)_2(PO_4)_2$

    ANSWER: c. $(NH_4)_3PO_4$

43. The name for the ClO_4^- ion is
    a. hypochlorite ion
    b. chlorite ion
    c. chlorate ion
    d. perchlorate ion
    e. chloroxide ion

    ANSWER: d. perchlorate ion

44. Which of the following is named incorrectly? What should its name be?
    a. $FeSO_4$; iron(II) sulfate
    b. $Sn_3(PO_4)_4$; tin(IV) phosphate
    c. K_3P; potassium phosphide
    d. $Fe(OH)_2$; iron(III) hydroxide
    e. All are correct.

    ANSWER: d. This answer is incorrect. It should be $Fe(OH)_2$; iron(II) hydroxide.

45. Potassium chlorate has the formula
    a. KCl
    b. $KClO$
    c. $KClO_2$
    d. $KClO_3$
    e. $KClO_4$

    ANSWER: d. $KClO_3$

46. Sodium chlorite has the formula
    a. $NaCl$
    b. $NaClO$
    c. $NaClO_2$
    d. $NaClO_3$
    e. $NaClO_4$

    ANSWER: c. $NaClO_2$

47. Which of the following represents the hypochlorite ion?
     a. ClO^-
     b. ClO_2^-
     c. ClO_3^-
     d. ClO_4^-
     e. Cl^-

     ANSWER: a. ClO^-

48. The formula for the compound formed from ammonium and sulfate ions is
     a. NH_4SO_4
     b. $(NH_4)_2SO_4$
     c. $NH_4(SO_4)_2$
     d. $(NH_4)_3SO_4$
     e. none of these

     ANSWER: b. $(NH_4)_2SO_4$

49. The name of the ClO_2^- ion is
     a. chlorite
     b. chloride
     c. hypochlorate
     d. perchlorate
     e. none of these

     ANSWER: a.   chlorite

50. The name for the compound NH_4F is _____.

     ANSWER: ammonium fluoride

51. The name for the compound $Sn(NO_3)_2$ is _____.

     ANSWER: tin(II) nitrate

52. The name for the compound Fe_2O_3 is _____.

     ANSWER: iron(III) oxide

53. The correct name for Cu_2O is
     a. copper oxide
     b. copper(I) oxide
     c. copper(II) oxide
     d. dicopper oxide
     e. dicopper monoxide

     ANSWER: b.   copper(I) oxide

54. The correct name for an aqueous solution of HBr is
    a. bromic acid
    b. hydrobromic acid
    c. hypobromic acid
    d. hydrobromous acid
    e. hypobromous acid

    ANSWER: b. hydrobromic acid

55. The name for the acid HNO_3 is _____.

    ANSWER: nitric acid

56. The correct name for $HClO_4(aq)$ is _____.

    ANSWER: perchloric acid

57. The name for the acid H_2SO_3 is
    a. sulfuric acid
    b. sulfurous acid
    c. hydrosulfuric acid
    d. hydrosulfurous acid
    e. sulfurite acid

    ANSWER: b. sulfurous acid

58. The name for the acid HNO_2 is
    a. nitrous acid
    b. nitric acid
    c. hydronitrous acid
    d. hydronitric acid
    e. hydrogen nitrite acid

    ANSWER: a. nitrous acid

59. The name for the acid H_2SO_4 is _____.

    ANSWER: sulfuric acid

60. The name for the acid $HC_2H_3O_2$ is _____.

    ANSWER: acetic acid

61. What is the name for HNO_3?
    a. nitrous acid
    b. nitric acid
    c. ammonia
    d. hydrocyanic acid
    e. hydrogen nitride

    ANSWER: b. nitric acid

62. Which of the following is the correct formula for bromic acid?
    a. $HBr(aq)$
    b. $HBrO(aq)$
    c. $HBrO_2(aq)$
    d. $HBrO_3(aq)$
    e. $HBrO_4(aq)$

    ANSWER: d. $HBrO_3(aq)$

63. Which of the following is named correctly?
    a. $HCl(aq)$; hypochlorous acid
    b. $(NH_4)_3PO_3$; ammonium phosphate
    c. $H_2SO_3(aq)$; sulfuric acid
    d. NH_3; ammonium ion
    e. $HNO_3(aq)$; nitric acid

    ANSWER: e. $HNO_3(aq)$; nitric acid

64. The correct formula for hypobromous acid is
    a. HBr
    b. $HBrO$
    c. $HBrO_2$
    d. $HBrO_3$
    e. $HBrO_4$

    ANSWER: b. $HBrO$

65. An aqueous solution of HF would be named
    a. hydrogen fluoride
    b. perfluoric acid
    c. hypofluorous acid
    d. hydrofluoric acid
    e. none of these

    ANSWER: d. hydrofluoric acid

66. Which of the following compounds is *not* named as an acid?
    a. $HCl(aq)$
    b. $HCN(aq)$
    c. $H_2SO_4(aq)$
    d. $HNO_3(aq)$
    e. $NH_3(aq)$

    ANSWER: e. $NH_3(aq)$

67. The name for K^+ is
    a. charged potassium
    b. elemental potassium
    c. potassium plus
    d. potassium
    e. potassium ion

    ANSWER: e.   potassium ion

68. The name for Ba_3P_2 is
    a. barium(II) phosphide
    b. tribarium diphosphide
    c. barium(I) phosphide
    d. barium phosphide
    e. barium phosphate

    ANSWER: d.   barium phosphide

69. The name for $HClO_3(aq)$ is
    a. chloric acid
    b. hydrogen chlorate
    c. perchloric acid
    d. hydrogen chlorate
    e. chlorous acid

    ANSWER: a.   chloric acid

70. The name for NH_4Br is
    a. nitrogen hydrogen bromide
    b. ammonium bromide
    c. ammonium(I) bromide
    d. nitrogen tetrahydrogen bromide
    e. none of these

    ANSWER: b.   ammonium bromide

71. The name for the Sr^{2+} species is
    a. strontium(II) ion
    b. strontium ion
    c. strontium(I) ion
    d. strontium
    e. distrontium ion

    ANSWER: b.   strontium ion

72. The name for CaO is
    a. monocalcium monoxide
    b. calcium(II) oxide
    c. calcium(I) oxide
    d. calcium oxide
    e. calcium monoxide

    ANSWER: d.   calcium oxide

73. The name for $NaHCO_3$ is
    a. sodium hydrogen carbonate (sodium bicarbonate)
    b. sodium carbonate
    c. sodium(I) hydrogen carbonate
    d. sodium(I) bicarbonate
    e. none of these

    ANSWER: a.   sodium hydrogen carbonate (sodium bicarbonate)

74. The name for Na_3N is
    a. sodium nitride
    b. trisodium nitride
    c. sodium(I) nitride
    d. sodium(III) nitride
    e. trisodium mononitride

    ANSWER: a.   sodium nitride

75. The name for $Al(OH)_3$ is
    a. aluminum(III) hydroxide
    b. aluminum trihydroxide
    c. aluminum hydroxide
    d. monaluminum trihydroxide
    e. aluminum(I) hydroxide

    ANSWER: c.   aluminum hydroxide

76. The name for $Ba(NO_3)_2$ is
    a. barium dinitrate
    b. barium(II) nitrate
    c. barium nitrite
    d. barium(I) nitrate
    e. barium nitrate

    ANSWER: e.   barium nitrate

77. The name for $HBrO(aq)$ is
    a. hydrogen bromous acid
    b. bromous acid
    c. bromic acid
    d. hypobromous acid
    e. perbromic acid

    ANSWER: d.   hypobromous acid

78. The name for $BaSO_4$ is _____.

    ANSWER: barium sulfate

79. The name for $FeCl_3$ is _____.

    ANSWER: iron(III) chloride

80. The name for PCl_5 is _____.

    ANSWER: phosphorus pentachloride

81. The name for N_2O is _____.

    ANSWER: dinitrogen monoxide

82. The name for $AgCl$ is _____.

    ANSWER: silver(I) chloride (commonly called silver chloride)

83. The name for NH_4Br is _____.

    ANSWER: ammonium bromide

84. The name for $CuCl_2$ is _____.

    ANSWER: copper(II) chloride

85. The name for SF_6 is _____.

    ANSWER: sulfur hexafluoride

86. The name for NO is _____.

    ANSWER: nitrogen monoxide

87. The name for PO_4^{3-} is _____.

    ANSWER: phosphate ion

88. The name for the compound Na_2SO_4 is _____.

    ANSWER: sodium sulfate

89. The name for the compound HgI_2 is _____.

    ANSWER: mercury(II) iodide

90. The name for the compound Hg_2Cl_2 is _____.

    ANSWER: mercury(I) chloride

91. The name for the compound FeO is _____.

    ANSWER: iron(II) oxide

92. The name for MnO_2 is _____.

    ANSWER: manganese(IV) oxide

93. The name for HCN*(aq)* is _____.

    ANSWER: hydrocyanic acid

94. The name for $MgCl_2$ is _____.

    ANSWER: magnesium chloride

95. The name for NH_4^+ is _____.

    ANSWER: ammonium ion

96. The name for the compound K_2O is _____.

    ANSWER: potassium oxide

97. The name for the compound BeH_2 is _____.

    ANSWER: beryllium hydride

98. The name for the compound $NaHCO_3$ is _____.

    ANSWER: sodium hydrogen carbonate (or sodium bicarbonate)

99. The name for the compound PI_3 is _____.

    ANSWER: phosphorus triiodide

100. The name for the compound $Na_2Cr_2O_7$ is _____.

    ANSWER: sodium dichromate

101. The name for the compound SnO_2 is _____.

    ANSWER: tin(IV) oxide

102. The name for $Mg(HCO_3)_2$ is _____.

    ANSWER: magnesium hydrogen carbonate

103. The name for $(NH_4)_2SO_4$ is _____.

    ANSWER: ammonium sulfate

104. The name for $H_2S(aq)$ is _____.

     ANSWER: hydrosulfuric acid

105. The name for $Co(OH)_2$ is _____.

     ANSWER: cobalt(II) hydroxide

106. The name for N_2S_3 is _____.

     ANSWER: dinitrogen trisulfide

107. The name for $CuBrO_2$ is _____.

     ANSWER: copper(I) bromite

108. The name for $SrHPO_4$ is _____.

     ANSWER: strontium hydrogen phosphate

109. The name for $Cu_3(PO_4)_2$ is _____.

     ANSWER: copper(II) phosphate

110. The name for $HCl(aq)$ is _____.

     ANSWER: hydrochloric acid

111. The name for $KC_2H_3O_2$ is _____.

     ANSWER: potassium acetate

112. The name for $HCN(g)$ is _____.

     ANSWER: hydrogen cyanide

113. The name for $NaCl$ is _____.

     ANSWER: sodium chloride

114. The name for $CuCl_2$ is _____.

     ANSWER: copper(II) chloride

115. The name for $Ba(NO_3)_2$ is _____.

     ANSWER: barium nitrate

116. The name for Ag_2CO_3 is _____.

     ANSWER: silver(I) carbonate (commonly called silver carbonate)

117.   The name for SiO_2 is _____.

ANSWER: silicon dioxide

118.   The name for $Mn(C_2H_3O_2)_2$ is _____.

ANSWER: manganese(II) acetate

119.   The name for $FeCl_3$ is _____.

ANSWER: iron(III) chloride

120.   The name for $Zn(OH)_2$ is _____.

ANSWER: zinc(II) hydroxide

CHAPTER 5b

Nomenclature

1. The correct formula for iron(III) oxide is
   a. Fe_3O_2
   b. FeO
   c. Fe_3O
   d. Fe_2O_3
   e. Fe_3O_3

   ANSWER: d. Fe_2O_3

2. The correct formula for sodium phosphate is
   a. Na_3PO_4
   b. $NaPO_4$
   c. Na_3PO_3
   d. Na_2PO_4
   e. Na_2PO_3

   ANSWER: a. Na_3PO_4

3. The correct formula for sodium bromide is
   a. Na_2Br
   b. SBr
   c. SBr_2
   d. $NaBr_2$
   e. $NaBr$

   ANSWER: e. $NaBr$

4. Which of the following is named correctly?
   a. $HCl(aq)$; hypochlorous acid
   b. $(NH_4)_3PO_3$; ammonium phosphate
   c. $H_2SO_3(aq)$; sulfuric acid
   d. $Cu(OH)_2$; copper(II) hydroxide
   e. $HNO_3(aq)$; nitrous acid

   ANSWER: d. $Cu(OH)_2$; copper(II) hydroxide

5. The correct formula for iron (III) nitrate is
    a. $Fe_2(NO_3)_2$
    b. $Fe(NO_3)_3$
    c. $Fe(NO_3)_2$
    d. $Fe_2(NO_3)_3$
    e. $Fe_3(NO_3)_2$

    ANSWER: b. $Fe(NO_3)_3$

6. What is the formula for copper(I) chloride?
    a. $CuCl_2$
    b. $CuCl$
    c. $CuClO$
    d. $CuClO_2$
    e. $CuClO_3$

    ANSWER: b. $CuCl$

7. What is the formula for manganese(IV) oxide?
    a. MgO_2
    b. Mn_4O
    c. MnO_2
    d. Mn_2O
    e. Mg_2O

    ANSWER: c. MnO_2

8. The correct formula for silicon dioxide is
    a. Si_2O_2
    b. Si_2O
    c. SiO_2
    d. SiO
    e. none of these

    ANSWER: c. SiO_2

9. The correct formula for nitrous acid is
    a. $H_2(NO_3)_3(aq)$
    b. $HNO_3(aq)$
    c. $H_2NO_2(aq)$
    d. $HNO_2(aq)$
    e. $H_2(NO_2)_2(aq)$

    ANSWER: d. $HNO_2(aq)$

10. The correct formula for sodium sulfite is
    a. Na_2SO_4
    b. Na_2SO_3
    c. $Na(SO_4)_2$
    d. $Na(SO_3)_2$
    e. $Na(SO_4)_3$

    ANSWER: b. Na_2SO_3

11. What is the formula for nickel(II) carbonate?
    a. NiC
    b. $NiCO_3$
    c. $Ni(CO_3)_2$
    d. Ni_2CO_3
    e. $Ni_2(CO_3)_3$

    ANSWER: b. $NiCO_3$

12. What is the formula for potassium sulfate?
    a. K_2SO_4
    b. K_2SO_3
    c. KSO_4
    d. KSO_3
    e. K_2S

    ANSWER: a. K_2SO_4

13. What is the formula for sulfur hexafluoride?
    a. S_3F
    b. SF_4
    c. S_6F
    d. SF_5
    e. SF_6

    ANSWER: e. SF_6

14. The correct formula for hydrocyanic acid is _____.

    ANSWER: HCN*(aq)*

15. Write the correct formula for barium phosphate.

    ANSWER: $Ba_3(PO_4)_2$

16. Write the correct formula for lithium hydride.

    ANSWER: LiH

17. Write the correct formula for tin(II) fluoride.

     ANSWER: SnF_2

18. Write the correct formula for sulfurous acid.

     ANSWER: H_2SO_3

19. Write the correct formula for silver(I) nitrate.

     ANSWER: $AgNO_3$

20. Write the correct formula for lead(IV) oxide.

     ANSWER: PbO_2

21. Write the correct formula for sulfuric acid.

     ANSWER: $H_2SO_4(aq)$

22. Write the correct formula for sodium nitrite.

     ANSWER: $NaNO_2$

23. Write the correct formula for beryllium hydride.

     ANSWER: BeH_2

24. Write the correct formula for calcium hydrogen carbonate.

     ANSWER: $Ca(HCO_3)_2$

25. Write the correct formula for iron(III) bromide.

     ANSWER: $FeBr_3$

26. Write the correct formula for ammonium dichromate.

     ANSWER: $(NH_4)_2Cr_2O_7$

27. Write the correct formula for dinitrogen pentoxide.

     ANSWER: N_2O_5

28. Write the correct formula for mercury(II) cyanide.

     ANSWER: $Hg(CN)_2$

29. Write the correct formula for acetic acid.

     ANSWER: $HC_2H_3O_2$

30. Write the correct formula for sodium hydride.

    ANSWER: NaH

31. Write the correct formula for iron(III) sulfide.

    ANSWER: Fe_2S_3

32. Write the correct formula for ammonium phosphate.

    ANSWER: $(NH_4)_3PO_4$

33. Give the formula for sodium bromide.

    ANSWER: NaBr

34. Give the formula for ammonium sulfate.

    ANSWER: $(NH_4)_2SO_4$

35. Give the formula for cobalt(II) chloride.

    ANSWER: $CoCl_2$

36. Give the formula for iron(III) oxide.

    ANSWER: Fe_2O_3

37. Give the formula for barium phosphate.

    ANSWER: $Ba_3(PO_4)_2$

38. Give the formula for iron(III) chloride.

    ANSWER: $FeCl_3$

39. Give the formula for magnesium hydride.

    ANSWER: MgH_2

40. Give the formula for calcium hydrogen carbonate.

    ANSWER: $Ca(HCO_3)_2$

41. Give the formula for aluminum chloride.

    ANSWER: $AlCl_3$

42. Give the formula for hydrosulfuric acid.

    ANSWER: H_2S

43. Give the formula for nitric acid.

    ANSWER: HNO_3

44. Give the formula for lithium sulfate.

    ANSWER: Li_2SO_4

45. Give the formula for dinitrogen pentoxide.

    ANSWER: N_2O_5

46. Give the formula for barium hydroxide.

    ANSWER: $Ba(OH)_2$

47. Give the formula for hypochlorous acid.

    ANSWER: $HClO$

48. Give the formula for ammonium cyanide.

    ANSWER: NH_4CN

49. Give the formula for potassium permanganate.

    ANSWER: $KMnO_4$

50. Give the formula for bromic acid.

    ANSWER: $HBrO_3$

51. Give the formula for sulfurous acid.

    ANSWER: $H_2SO_3(aq)$

52. Give the formula for cobalt(II) nitrate.

    ANSWER: $Co(NO_3)_2$

53. Give the formula for hydroiodic acid.

    ANSWER: $HI(aq)$

54. Give the formula for lithium sulfide.

    ANSWER: Li_2S

55. Give the formula for sulfur dioxide.

    ANSWER: SO_2

56. Give the formula for manganese(II) chloride.

    ANSWER: $MnCl_2$

57. Give the formula for hydrogen peroxide.

    ANSWER: H_2O_2

58. Give the formula for carbonic acid.

    ANSWER: H_2CO_3

59. Give the formula for dinitrogen tetroxide.

    ANSWER: N_2O_4

60. Give the formula for iron(III) oxide.

    ANSWER: Fe_2O_3

61. Give the formula for calcium carbonate.

    ANSWER: $CaCO_3$

62. Give the formula for rubidium hydrogen carbonate.

    ANSWER: $RbHCO_3$

63. Give the formula for mercury(II) chloride.

    ANSWER: $HgCl_2$

64. Give the formula for aluminum oxide.

    ANSWER: Al_2O_3

65. Give the formula for dinitrogen pentoxide.

    ANSWER: N_2O_5

66. Give the formula for perchloric acid.

    ANSWER: $HClO_4$

67. Give the formula for silver(I) cyanide.

    ANSWER: AgCN

68. Give the formula for magnesium phosphate.

    ANSWER: $Mg_3(PO_4)_2$

69. Give the formula for chromium(III) iodide.

    ANSWER: CrI_3

70. Give the formula for sulfurous acid.

    ANSWER: H_2SO_3

71. Give the formula for mercury(II) sulfide.

    ANSWER: HgS

72. Give the formula for sodium hydroxide.

    ANSWER: NaOH

73. Give the formula for silicon tetrafluoride.

    ANSWER: SiF_4

74. Give the formula for potassium permanganate.

    ANSWER: $KMnO_4$

75. Give the formula for sodium chromate.

    ANSWER: Na_2CrO_4

76. Give the formula for potassium chlorate.

    ANSWER: $KClO_3$

77. Give the formula for titanium(IV) chloride.

    ANSWER: $TiCl_4$

78. Give the formula for barium peroxide.

    ANSWER: BaO_2

79. What is the formula for calcium hydroxide?
    a. CaOH
    b. $Ca(OH)_2$
    c. $Ca(OH)_3$
    d. $Ca_2(OH)_3$
    e. none of these

    ANSWER: b. $Ca(OH)_2$

80. Give the formula for sodium nitrate.

    ANSWER: $NaNO_3$

81. Give the formula for iron(III) sulfate.

    ANSWER: $Fe_2(SO_4)_3$

82. Give the formula for potassium iodide.

    ANSWER: KI

83. Give the formula for lead(II) fluoride.

    ANSWER: PbF_2

84. Give the formula for barium hydroxide.

    ANSWER: $Ba(OH)_2$

85. Give the formula for hydroiodic acid.

    ANSWER: HI*(aq)*

86. Give the name for H_3PO_4*(aq)*.

    ANSWER: phosphoric acid

87. Give the name for $LiC_2H_3O_2$

    ANSWER: lithium acetate

88. Give the name for K_2CO_3.

    ANSWER: potassium carbonate

89. Give the name for $Hg(HCO_3)_2$

    ANSWER: mercury(II) hydrogen carbonate

90. Give the name for $Fe(OH)_3$.

    ANSWER: iron(III) hydroxide

91.   Give the name for CO_2.

ANSWER: carbon dioxide

92.   Give the name for $Al(OH)_3$.

ANSWER: aluminum hydroxide

93.   Give the name for SnS_2.

ANSWER: tin(IV) sulfide

94.   Give the formula for ammonia.

ANSWER: NH_3.

95.   Give the formula for dinitrogen monoxide.

ANSWER: N_2O

96.   Give the formula for carbon monoxide.

ANSWER: CO

97.   Give the name for SiO_2.

ANSWER: silicon dioxide

98.   Give the name for NaI.

ANSWER: sodium iodide

99.   Give the name for K_2S.

ANSWER: potassium sulfide

CHAPTER 6

Chemical Reactions: An Introduction

1. All the following are clues that a chemical reaction has taken place *except*
   a. a color change
   b. a solid forms
   c. the reactant is smaller
   d. bubbles form
   e. a flame occurs

   ANSWER: c.   the reactant is smaller

2. The equation $N_2 + 3H_2 \rightarrow 2NH_3$ means that 1 g of N_2 reacts with 3 g of H_2 to form 2 g of NH_3.
   a. True
   b. False

   ANSWER: b.   False

3. You are asked to balance the chemical equation $H_2 + O_2 \rightarrow H_2O$. How many of the following ways are correct ways to balance this equation?
   I. $2H_2 + O_2 \rightarrow 2H_2O$.
   II. $H_2 + \frac{1}{2}O_2 \rightarrow H_2O$.
   III. $4H_2 + 2O_2 \rightarrow 4H_2O$.
   IV. $H_2 + O_2 \rightarrow H_2O_2$.
   a. 0
   b. 1
   c. 2
   d. 3
   e. 4

   ANSWER: d. 3

4. Aluminum oxide solid reacts with gaseous carbon monoxide to produce aluminum metal and carbon dioxide gas. Write the balanced equation for this reaction.

   ANSWER: $Al_2O_3(s) + 3CO(g) \rightarrow 2Al(s) + 3CO_2(g)$

5. Sodium metal reacts with water to produce aqueous sodium hydroxide and hydrogen gas. Write the balanced equation for this reaction.

   ANSWER: $2Na(g) + 2H_2O(l) \rightarrow 2NaOH(aq) + H_2(g)$

76 *Chapter 6*

When the following equations are balanced using the smallest possible integers, what is the number in front of the underlined substance in each case?

6. $C_3H_8(g) + \underline{O_2(g)} \rightarrow CO_2(g) + H_2O(g)$

    a. 2
    b. 3
    c. 4
    d. 5
    e. 6

    ANSWER: d. 5

7. $Mg(s) + HCl(aq) \rightarrow MgCl_2(aq) + \underline{H_2(g)}$

    a. 1
    b. 2
    c. 3
    d. 4
    e. 5

    ANSWER: a. 1

8. $H_3PO_4(aq) + Ca(OH)_2(aq) \rightarrow Ca_3(PO_4)_2(aq) + \underline{H_2O(l)}$

    a. 2
    b. 3
    c. 4
    d. 5
    e. 6

    ANSWER: e. 6

9. $Na(s) + \underline{H_2O(l)} \rightarrow NaOH(aq) + H_2(g)$

    a. 2
    b. 3
    c. 4
    d. 5
    e. 6

    ANSWER: a. 2

10. $\underline{MgO(s)} \rightarrow Mg(s) + O_2(g)$

    a. 2
    b. 3
    c. 4
    d. 5
    e. 6

    ANSWER: a. 2

11. $Al(s) + \underline{O_2(g)} \rightarrow Al_2O_3(s)$

    a.   2
    b.   3
    c.   4
    d.   5
    e.   6

    ANSWER: b.   3

12. $HCl(aq) + Mg(OH)_2(aq) \rightarrow \underline{MgCl_2(aq)} + H_2O(l)$

    a.   5
    b.   4
    c.   3
    d.   2
    e.   1

    ANSWER: e.   1

13. $SO_2(g) + O_2(g) \rightarrow \underline{SO_3(g)}$

    a.   5
    b.   4
    c.   3
    d.   2
    e.   1

    ANSWER: d.   2

14. $Br_2(l) + KI(aq) \rightarrow \underline{I_2(aq)} + KBr(aq)$

    a.   1
    b.   2
    c.   3
    d.   4
    e.   6

    ANSWER: b.   2

15.   The sum of the coefficients when the following equation is balanced is
    $BaSO_4 + K_3PO_4 \rightarrow Ba_3(PO_4)_2 + K_2SO_4$

    a.   4
    b.   7
    c.   8
    d.   9
    e.   11

    ANSWER: d.   9

16. Which of the following statements concerning balancing equations is false?
    a. There must always be a coefficient of 1 in an equation balanced in standard form.
    b. The ratio of coefficients is much more meaningful than an individual coefficient in a balanced equation.
    c. The total number of atoms must be the same on the reactants side and the product side of the balanced equation.
    d. When there are two products, the order in which they are written does not matter.
    e. At least two of the above statements (a-d) are false.

    ANSWER: a.   1

17. What is the sum of the coefficients of the following equation when it is balanced using smallest whole number integers? $Na_2S_2O_3 + I_2 \rightarrow Na_2S_4O_6 + NaI$
    a. 4
    b. 5
    c. 6
    d. 7
    e. 8

    ANSWER: c.   6

18. Consider the reaction represented by the unbalanced equation $NH_3 + O_2 \rightarrow NO + H_2O$.
    For every 1.00 mol of NH_3 that reacts, _____ mol of O_2 is required.
    a. 1
    b. 1.25
    c. 4.00
    d. 5.00
    e. none of these

    ANSWER: b.   1.25

19. Consider a reaction given by the equation $aA + bB \rightarrow cC + dD$. In this equation A, B, C, D represent chemicals, and a, b, c, d represent coefficients in the balanced equation. For a given reaction, how many values are there for the quantity "c/d"?
    a. 1
    b. 2
    c. 3
    d. 4
    e. an infinite number

    ANSWER: a.   1

20. Choose the response that best answers the question "Why do we not change subscripts when balancing a chemical equation?"
    a. There's no real reason not to, it's just something that is not done.
    b. We never change subscripts because atoms are neither created nor destroyed. This is part of the law of conservation of matter.
    c. It's fine to do once in awhile, but don't make a habit of it.
    d. We can change subscripts. We can't change coefficients.
    e. Compounds composed of the same elements and having different subscripts are different substances.

    ANSWER: e.   Compounds composed of the same elements and having different subscripts are different substances.

When the following equations are balanced using the smallest possible integers, what is the number in front of the underlined substance in each case?

21. $Pb(s) + \underline{AgNO_3}(aq) \rightarrow Pb(NO_3)_2 + Ag(s)$

   a. 1
   b. 2
   c. 4
   d. 5
   e. 6

   ANSWER: b. 2

22. $S(s) + O_2(g) \rightarrow \underline{SO_2}(g)$

   a. 1
   b. 2
   c. 3
   d. 4
   e. 5

   ANSWER: a. 1

23. $N_2(g) + \underline{O_2}(g) \rightarrow N_2O_3$

   a. 1
   b. 2
   c. 3
   d. 4
   e. 6

   ANSWER: c. 3

24. $Al_2O_3(s) + H_2SO_4(aq) \rightarrow Al_2(SO_4)_3(aq) + \underline{H_2O}(l)$

   a. 1
   b. 2
   c. 3
   d. 6
   e. 9

   ANSWER: c. 3

25. $Sr(s) + P_4(s) \rightarrow \underline{Sr_3P_2}(s)$

   a. 2
   b. 3
   c. 6
   d. 12
   e. 18

   ANSWER: c. 6

26. $C_2H_6(g) + \underline{O_2(g)} \rightarrow CO_2(g) + H_2O(g)$

    a. 4
    b. 7
    c. 8
    d. 10
    e. 14

    ANSWER: b. 7

27. $Li(s) + O_2(g) \rightarrow \underline{Li_2O(s)}$

    a. 1
    b. 2
    c. 3
    d. 4
    e. 5

    ANSWER: b. 2

28. Balance the equation
    $NaBH_4 + BF_3 \rightarrow NaBF_4 + B_2H_6$

    ANSWER: $3NaBH_4 + 4BF_3 \rightarrow 3NaBF_4 + 2B_2H_6$

29. Balance the equation
    $MgCl_2 + K_3PO_4 \rightarrow Mg_3(PO_4)_2 + KCl$

    ANSWER: $3MgCl_2 + 2K_3PO_4 \rightarrow Mg_3(PO_4)_2 + 6KCl$

30. Balance the equation
    $C_6H_{14} + O_2 \rightarrow CO_2 + H_2O$

    ANSWER: $2C_6H_{14} + 19O_2 \rightarrow 12CO_2 + 14H_2O$

31. Balance the equation
    $As_2O_3(s) + Ca(OH)_2(aq) \rightarrow Ca_3(AsO_4)_2(s) + H_2O(l)$

    ANSWER: $As_2O_3(s) + 3Ca(OH)_2(aq) \rightarrow Ca_3(AsO_4)_2(s) + 3H_2O(l)$

32. Balance the equation
    $Mg(OH)_2(aq) + 2HBr(aq) \rightarrow MgBr_2(aq) + 2H_2O(l)$

    ANSWER: $Mg(OH)_2(aq) + 2HBr(aq) \rightarrow MgBr_2(aq) + 2H_2O(l)$

33. Balance the equation
    $Sb(s) + O_2(g) \rightarrow Sb_2O_3(s)$

    ANSWER: $4Sb(s) + 3O_2(g) \rightarrow 2Sb_2O_3(s)$

When the following equations are balanced using the smallest possible integers, what is the number in front of the underlined substance in each case?

34. $As(OH)_3(s) + H_2SO_4(aq) \rightarrow As_2(SO_4)_3(aq) + \underline{H_2O}(l)$

    a. 1
    b. 2
    c. 4
    d. 6
    e. 12

    ANSWER: d. 6

35. $\underline{Al}(s) + O_2(g) \rightarrow Al_2O_3(s)$

    a. 1
    b. 2
    c. 4
    d. 6
    e. 12

    ANSWER: c. 4

36. $FeCl_2(aq) + Ag_3PO_4(aq) \rightarrow Fe_3(PO_4)_2(aq) + \underline{AgCl}(s)$

    a. 1
    b. 2
    c. 4
    d. 6
    e. 12

    ANSWER: d. 6

37. Balance the equation
    $Mg(s) + H_3PO_4(aq) \rightarrow Mg_3(PO_4)_2(aq) + H_2(g)$

    ANSWER: $3Mg(s) + 2H_3PO_4(aq) \rightarrow Mg_3(PO_4)_2(aq) + 3H_2(g)$

38. Balance the equation
    $KClO_3(s) \rightarrow KCl(s) + O_2(g)$

    ANSWER: $2KClO_3(s) \rightarrow 2KCl(s) + 3O_2(g)$

39. When table sugar is burned in air, carbon dioxide and water vapor are products as shown by the following unbalanced chemical equation
    $$C_{12}H_{22}O_{11}(s) + O_2(g) \rightarrow CO_2(g) + H_2O(g)$$
    How many moles of oxygen are required to react completely with 1.0 mol of sugar?
    a. 12
    b. 17.5
    c. 24
    d. 35
    e. none of these

    ANSWER: a. 12

40. When ethane (C_2H_6) is reacted with oxygen in the air, the products are carbon dioxide and water. This process requires _____ mol of oxygen for every one mole of ethane.
    a. 1
    b. 2.5
    c. 3.5
    d. 7
    e. none of these

    ANSWER: c.   3.5

41. Write and balance the equation showing the reaction between iron(III) oxide and carbon monoxide to form iron metal and carbon dioxide.

    ANSWER: $Fe_2O_3(s) +3CO(g) \rightarrow 2Fe(s) +3CO_2(s)$

42. Write and balance the equation showing the reaction between nitrogen monoxide gas and hydrogen gas to form nitrogen gas and water vapor.

    ANSWER: $2NO(g) +2H_2(g) \rightarrow N_2(g) +2H_2O(g)$

43. Write and balance the equation showing the reaction between calcium metal and water to form aqueous calcium hydroxide and hydrogen gas.

    ANSWER: $Ca(s) +2H_2O(l) \rightarrow Ca(OH)_2(aq) +H_2(g)$

44. Write and balance the equation showing the reaction between iron(III) oxide and carbon monoxide to form iron(II) oxide and carbon dioxide.

    ANSWER: $Fe_2O_3(s) +CO(g) \rightarrow 2FeO(s) +CO_2(s)$

45. Write and balance the equation showing the reaction between copper metal and aqueous sulfuric acid to form aqueous copper(II) sulfate, sulfur dioxide gas, and water.

    ANSWER: $Cu(s) +2H_2SO_4(aq) \rightarrow CuSO_4(aq) + SO_2(g) + 2H_2O(l)$

46. Write and balance the equation showing the reaction between gaseous hydrogen sulfide and oxygen to form sulfur dioxide gas and water vapor.

    ANSWER: $2H_2S(g) +3O_2(g) \rightarrow 2SO_2(g) + 2H_2O(g)$

47. Balance the equation
    $H_2O_2(l) \rightarrow H_2O(l) + O_2(g)$

    ANSWER: $2H_2O_2(l) \rightarrow 2H_2O(l) + O_2(g)$

48. Balance the equation
    $Mg(s) + O_2(g) \rightarrow MgO(s)$

    ANSWER: $2Mg(s) + O_2(g) \rightarrow 2MgO(s)$

49. Balance the equation
$BaCl_2(aq) + H_2SO_4(aq) \rightarrow BaSO_4(s) + HCl(aq)$

ANSWER: $BaCl_2(aq) + H_2SO_4(aq) \rightarrow BaSO_4(s) + 2HCl(aq)$

When the following equations are balanced using the smallest possible integers, what is the number in front of the underlined substance in each case?

50. $\underline{Sb(s)} + O_2(g) \rightarrow Sb_2O_5(s)$
   a. 1
   b. 2
   c. 4
   d. 6
   e. 12

   ANSWER: c. 4

51. $C_4H_{10}(g) + O_2(g) \rightarrow \underline{CO_2(g)} + H_2O(g)$
   a. 2
   b. 4
   c. 6
   d. 8
   e. 10

   ANSWER: d. 8

52. $CH_3OH(l) + O_2(g) \rightarrow CO_2(g) + \underline{H_2O(g)}$
   a. 1
   b. 2
   c. 4
   d. 6
   e. 12

   ANSWER: c. 4

53. Balance the equation
$Pb(NO_3)_2(aq) + K_2CrO_4(aq) \rightarrow PbCrO_4(s) + KNO_3(aq)$

ANSWER: $Pb(NO_3)_2(aq) + K_2CrO_4(aq) \rightarrow PbCrO_4(s) + 2KNO_3(aq)$

54. Balance the equation
$(NH_4)_2Cr_2O_7(s) \rightarrow N_2(g) + H_2O(g) + Cr_2O_3(s)$

ANSWER: $(NH_4)_2Cr_2O_7(s) \rightarrow N_2(g) + 4H_2O(g) + Cr_2O_3(s)$

55. When the equation $Cr_2S_3 + HCl \rightarrow CrCl_3 + H_2S$ is balanced in standard form, one of the terms in the balanced equation is
    a. $3HCl$
    b. $CrCl_3$
    c. $3H_2S$
    d. $2Cr_2S_3$

    ANSWER: c. $3H_2S$

56. Balance the equation for the reaction of calcium metal with oxygen gas to produce solid calcium oxide.

    ANSWER: $2Ca(s) + O_2(g) \rightarrow 2CaO(s)$

57. Balance the equation for the reaction of aluminum metal with solid iodine to form solid aluminum iodide.

    ANSWER: $2Al(s) + 3I_2(s) \rightarrow 2AlI_3(s)$

58. Balance the equation for the reaction of potassium metal with water to form potassium hydroxide and hydrogen gas.

    ANSWER: $2K(s) + H_2O(l) \rightarrow 2KOH(aq) + H_2(g)$

59. Balance the equation
    $KI(aq) + Cl_2(g) \rightarrow KCl(aq) + I_2(s)$

    ANSWER: $2KI(aq) + Cl_2(g) \rightarrow 2KCl(aq) + I_2(s)$

60. Balance the equation
    $Zn(s) + HCl(aq) \rightarrow H_2(g) + ZnCl_2(aq)$

    ANSWER: $Zn(s) + 2HCl(aq) \rightarrow H_2(g) + ZnCl_2(aq)$

61. Balance the equation
    $Na(s) + H_2O(l) \rightarrow H_2(g) + NaOH(aq)$

    ANSWER: $2Na(s) + 2H_2O(l) \rightarrow H_2(g) + 2NaOH(aq)$

62. Balance the equation
    $Ca(s) + O_2(g) \rightarrow CaO(s)$

    ANSWER: $2Ca(s) + O_2(g) \rightarrow 2CaO(s)$

63. Balance the equation
    $Mg_3N_2(s) + H_2O(l) \rightarrow NH_3(g) + Mg(OH)_2(s)$

    ANSWER: $Mg_3N_2(s) + 6H_2O(l) \rightarrow 2NH_3(g) + 3Mg(OH)_2(s)$

64. Balance the equation
     $AgNO_3(aq) + CaCl_2(aq) \rightarrow AgCl(s) + Ca(NO_3)_2(aq)$

     ANSWER: $2AgNO_3(aq) + CaCl_2(aq) \rightarrow 2AgCl(s) + Ca(NO_3)_2(aq)$

65. Balance the equation
     $CaCO_3(s) + HCl(aq) \rightarrow H_2O(l) + CO_2(g) + CaCl_2(aq)$

     ANSWER: $CaCO_3(s) + 2HCl(aq) \rightarrow H_2O(l) + CO_2(g) + CaCl_2(aq)$

66. Balance the equation
     $PBr_3(g) + H_2O(l) \rightarrow H_3PO_3(aq) + HBr(aq)$

     ANSWER: $PBr_3(g) + 3H_2O(l) \rightarrow H_3PO_3(aq) + 3HBr(aq)$

67. Balance the equation
     $Na_2O(s) + H_2O(l) \rightarrow NaOH(aq)$

     ANSWER: $Na_2O(s) + H_2O(l) \rightarrow 2NaOH(aq)$

68. Balance the equation
     $C_3H_7OH(l) + O_2(g) \rightarrow CO_2(g) + H_2O(g)$

     ANSWER: $2C_3H_7OH(l) + 9O_2(g) \rightarrow 6CO_2(g) + 8H_2O(g)$

69. Balance the equation
     $Fe(NO_3)_2(aq) + H_2S(g) \rightarrow FeS(s) + HNO_3(aq)$

     ANSWER: $Fe(NO_3)_2(aq) + H_2S(g) \rightarrow FeS(s) + 2HNO_3(aq)$

70. Balance the equation
     $C_8H_{18}(l) + O_2(g) \rightarrow CO_2(g) + H_2O(g)$

     ANSWER: $2C_8H_{18}(l) + 25O_2(g) \rightarrow 16CO_2(g) + 18H_2O(g)$

71. Balance the equation
     $K(s) + H_2O(l) \rightarrow KOH(aq) + H_2(g)$

     ANSWER: $2K(s) + 2H_2O(l) \rightarrow 2KOH(aq) + H_2(g)$

72. Balance the equation
     $NH_3(g) + O_2(g) \rightarrow NO(g) + H_2O(l)$

     ANSWER: $4NH_3(g) + 5O_2(g) \rightarrow 4NO(g) + 6H_2O(l)$

73. Balance the equation
     $HCl(aq) + Ca(OH)_2(aq) \rightarrow CaCl_2(aq) + H_2O(l)$

     ANSWER: $2HCl(aq) + Ca(OH)_2(aq) \rightarrow CaCl_2(aq) + 2H_2O(l)$

74. Balance the equation
    $$Al(s) + S(s) \rightarrow Al_2S_3(s)$$

    ANSWER: $2Al(s) + 3S(s) \rightarrow Al_2S_3(s)$

75. When the equation $Si(s) + HF(aq) \rightarrow SiF_4(g) + H_2(g)$ is balanced, what is the coefficient for HF?
    a. 0
    b. 1
    c. 2
    d. 3
    e. 4

    ANSWER: e. 4

76. Balance the equation
    $$Zn(s) + H_3PO_4(aq) \rightarrow Zn_3(PO_4)_2(s) + H_2(g)$$

    ANSWER: $3Zn(s) + 2H_3PO_4(aq) \rightarrow Zn_3(PO_4)_2(s) + 3H_2(g)$

CHAPTER 7

Reactions in Aqueous Solutions

1. The most common factors that cause chemical reactions to occur are all the following *except*
   a. formation of a solid
   b. formation of a gas
   c. formation of water
   d. transfer of electrons
   e. a decrease in temperature

   ANSWER: e.   a decrease in temperature

2. A substance that, when dissolved in water, produces a solution that conducts electric current very efficiently is called
   a. a strong electrolyte
   b. a weak electrolyte
   c. a strong ion
   d. an electrical solute
   e. none of these

   ANSWER: a.   a strong electrolyte

3. When a precipitation reaction occurs, the ions that do *not* form the precipitate
   a. evaporate
   b. are cations only
   c. form a second insoluble compound in the solution
   d. are left dissolved in the solution
   e. none of these

   ANSWER: d.   are left dissolved in the solution

Write and balance molecular equations for the following reactions between aqueous solutions. You will need to decide on the formulas and phases of the products in each of the cases.

4. An aqueous solution of barium hydroxide is mixed with an aqueous solution of sulfuric acid.

   ANSWER: $Ba(OH)_2(aq) + H_2SO_4(aq) \rightarrow BaSO_4(s) + 2H_2O(l)$

5. An aqueous solution of potassium chloride is mixed with an aqueous solution of sodium sulfate.

   ANSWER: $2KCl(aq) + Na_2SO_4(aq) \rightarrow K_2SO_4(aq) + 2NaCl(aq)$

6. An aqueous solution of barium nitrate is mixed with an aqueous solution of potassium phosphate.

   ANSWER: $3Ba(NO_3)_2(aq) + 2K_3PO_4(aq) \rightarrow Ba_3(PO_4)_2(s) + 6KNO_3(aq)$

7. An aqueous solution of silver nitrate is mixed with an aqueous solution of potassium chromate.

   ANSWER: $2AgNO_3(aq) + K_2CrO_4(aq) \rightarrow Ag_2CrO_4(s) + 2KNO_3(aq)$

8. An aqueous solution of lead(II) nitrate is mixed with an aqueous solution of sodium iodide.

   ANSWER: $Pb(NO_3)_2(aq) + 2NaI(aq) \rightarrow PbI_2(s) + 2NaNO_3(aq)$

9. An aqueous solution of silver nitrate is mixed with an aqueous solution of potassium carbonate.

   ANSWER: $2AgNO_3(aq) + K_2CO_3(aq) \rightarrow Ag_2CO_3(s) + 2KNO_3(aq)$

10. An aqueous solution of copper(II) nitrate is mixed with an aqueous solution of sodium hydroxide.

    ANSWER: $Cu(NO_3)_2(aq) + 2NaOH(aq) \rightarrow Cu(OH)_2(s) + 2NaNO_3(aq)$

11. An aqueous solution of calcium nitrate is mixed with an aqueous solution of sodium phosphate.

    ANSWER: $3Ca(NO_3)_2(aq) + 2Na_3PO_4(aq) \rightarrow Ca_3(PO_4)_2(s) + 6NaNO_3(aq)$

12. An aqueous solution of lead(II) nitrate is mixed with an aqueous solution of potassium chloride.

    ANSWER: $Pb(NO_3)_2(aq) + 2KCl(aq) \rightarrow PbCl_2(s) + 2KNO_3(aq)$

An aqueous solution of barium nitrate is allowed to react with an aqueous solution of sodium sulfate.

13. Identify the solid in the balanced equation.
    a. $NaNO_3$
    b. $BaSO_4$
    c. $NaNO_3$
    d. $BaSO_2$
    e. none of these

    ANSWER: b. $BaSO_4$

14. What is the coefficient of the solid in the balanced equation?

    ANSWER: one

15. The net ionic reaction for the reaction between aqueous lead nitrate and aqueous potassium iodide is
    a. $Pb(NO_3)_2(aq) + KI(aq) \rightarrow PbI_2(s) + KNO_3(aq)$
    b. $Pb^{2+}(aq) + NO_3^-(aq) + K^+(aq) + I^-(aq) \rightarrow Pb^{2+}(aq) + I^-(aq) + K^+(aq) + NO_3^-(aq)$
    c. $Pb^{2+}(aq) + 2NO_3^-(aq) + 2K^+(aq) + I^-(aq) \rightarrow PbI_2(s) + 2K^+(aq) + 2NO_3^-(aq)$
    d. $Pb^{2+}(aq) + 2I^-(aq) \rightarrow PbI_2(s)$
    e. none of these

    ANSWER: d. $Pb^{2+}(aq) + 2I^-(aq) \rightarrow PbI_2(s)$

16. What is the correct balanced complete ionic equation for the reaction of lead(II) nitrate with potassium chloride?
    a. $Pb(NO_3)_2(aq) + KCl(aq) \rightarrow PbCl_2(s) + KNO_3(aq)$
    b. $Pb^{2+}(aq) + 2NO_3^-(aq) + 2K^+(aq) + 2Cl^-(aq) \rightarrow Pb^{2+}(aq) + 2Cl^-(aq) + 2 K^+(aq) + 2NO_3^-(aq)$
    c. $Pb^{2+}(aq) + 2NO_3^-(aq) + 2K^+(aq) + 2Cl^-(aq) \rightarrow Pb^{2+}(aq) + 2Cl^-(aq) + 2KNO_3(s)$
    d. $Pb^{2+}(aq) + 2NO_3^-(aq) + K^+(aq) + Cl^-(aq) \rightarrow PbCl_2(s) + K^+(aq) + NO_3^-(aq)$
    e. $Pb^{2+}(aq) + 2NO_3^-(aq) + 2K^+(aq) + 2Cl^-(aq) \rightarrow PbCl_2(s) + 2 K^+(aq) + 2NO_3^-(aq)$

    ANSWER: e. $Pb^{2+}(aq) + 2NO_3^-(aq) + 2K^+(aq) + 2Cl^-(aq) \rightarrow PbCl_2(s) + 2 K^+(aq) + 2NO_3^-(aq)$

17. $AgNO_3(aq) + NH_4Cl(aq) \rightarrow$
    a. $Ag_2Cl(s)$
    b. $AgCl(s)$
    c. $NH_4NO_3(s)$
    d. $AgNH_4(s)$
    e. none of these

    ANSWER: b. $AgCl(s)$

18. Which of the following "hardens" tooth enamel?
    a. F_2
    b. F^-
    c. $Ca(OH)_2$
    d. $FeCl_3$
    e. Cl^-

    ANSWER: b. F^-

19. Write the balanced molecular equation for the reaction between aqueous solutions of lithium phosphate and sodium hydroxide.

    ANSWER: $Li_3PO_4(aq) + 2NaOH(aq) \rightarrow Na_3PO_4(aq) + 3LiOH(aq)$

20. Complete and write the balanced molecular equation for the following: An aqueous solution of magnesium chloride is added to an aqueous solution of silver nitrate.

    ANSWER: $MgCl_2(aq) + 2AgNO_3(aq) \rightarrow 2AgCl(s) + Mg(NO_3)_2(aq)$

90 *Chapter 7*

21. The _____ equation contains only those substances directly involved in reactions in aqueous solutions.
    a. molecular
    b. complete ionic
    c. net ionic
    d. reduced ionic
    e. ionic

    ANSWER: c.   net ionic

Aqueous solutions of sodium sulfide and silver nitrate are mixed to form solid silver sulfide and aqueous sodium nitrate.

22. The molecular equation contains which one of the following terms (when balanced in standard form)?
    a. $AgNO_3(aq)$
    b. $Ag_2S(aq)$
    c. $Na_2S(aq)$
    d. $2NaNO_3(s)$
    e. $2Ag_2S(s)$

    ANSWER: c. $Na_2S(aq)$

23. The complete ionic equation contains which of the following species (when balanced in standard form)?
    a. $2Na^+(aq)$
    b. $2S^{2-}(aq)$
    c. $Ag^+(aq)$
    d. $NO_3^-(aq)$
    e. $3NO_3^-(aq)$

    ANSWER: a. $2Na^+(aq)$

24. The net ionic equation contains which of the following species (when balanced in standard form)?
    a. $Ag^+(aq)$
    b. $S^{2-}(aq)$
    c. $NO_3^-(aq)$
    d. $2NO_3^-(aq)$
    e. $2Na^+(aq)$

    ANSWER: b. $S^{2-}(aq)$

25. Write the molecular equation, the complete ionic equation, and the net ionic equation for the following reaction: Aqueous solutions of potassium sulfate and barium chloride are mixed to form solid barium sulfate and aqueous potassium chloride.

    ANSWER:

    $K_2SO_4(aq) + BaCl_2(aq) \rightarrow BaSO_4(s) + 2KCl(aq)$

    $2K^+(aq) + SO_4^{2-}(aq) + Ba^{2+}(aq) + 2Cl^-(aq) \rightarrow BaSO_4(s) + 2K^+(aq) + 2Cl^-(aq)$

    $Ba^{2+}(aq) + SO_4^{2-}(aq) \rightarrow BaSO_4(s)$

26. Write the molecular equation, the complete ionic equation, and the net ionic equation for the following reaction: Aqueous solutions of copper(II) nitrate and sodium hydroxide are mixed to form solid copper(II) hydroxide and aqueous sodium nitrate.

    ANSWER:

    $Cu(NO_3)_2(aq) + 2NaOH(aq) \rightarrow Cu(OH)_2(s) + 2NaNO_3(aq)$

    $Cu^{2+}(aq) + 2NO_3^-(aq) + 2Na^+(aq) + 2OH^-(aq) \rightarrow Cu(OH)_2(s) + 2Na^+(aq) + 2NO_3^-(aq)$

    $Cu^{2+}(aq) + 2OH^-(aq) \rightarrow Cu(OH)_2(s)$

27. The scientist who discovered the essential nature of acids through solution conductivity studies is
    a. Priestly
    b. Boyle
    c. Einstein
    d. Mendeleev
    e. Arrhenius

    ANSWER: e. Arrhenius

28. When the following equation is balanced, what is the coefficient for H_2O?
    $Ca(OH)_2(aq) + H_3PO_4(aq) \rightarrow Ca_3(PO_4)_2(s) + H_2O(l)$
    a. 2
    b. 3
    c. 4
    d. 6
    e. 8

    ANSWER: d. 6

29. When the equation $Al(OH)_3(s) + H_2SO_4(aq) \rightarrow$ is completed and balanced in standard form, one of the terms in the balanced equation is
    a. $AlSO_4(s)$
    b. $3H_2O(l)$
    c. $6H_2SO_4(aq)$
    d. $Al_2(SO_4)_3(aq)$
    e. $2H_2SO_4(aq)$

    ANSWER: d. $Al_2(SO_4)_3(aq)$

30. When an acid reacts with a base, which product always forms?
    a. hydrogen
    b. carbon dioxide
    c. water
    d. hydrogen and carbon dioxide
    e. none of these

    ANSWER: c. water

31. Which of the following is a salt?
    a. HCl
    b. SO_3
    c. HNO_2
    d. $MgCl_2$
    e. none of these

    ANSWER: d. $MgCl_2$

32. When the equation $NaOH(aq) + H_2SO_4(aq) \rightarrow$ is completed and balanced in standard form, one of the terms in the balanced equation is
    a. $NaSO_4(aq)$
    b. $2H_2O(l)$
    c. $H_2OH(l)$
    d. $2Na_2SO_4(aq)$

    ANSWER: b. $2H_2O(l)$

33. Balance the complete ionic equation when aqueous barium hydroxide reacts with aqueous hydrochloric acid.

    ANSWER: $Ba^{2+}(aq) + 2OH^-(aq) + 2H^+(aq) + 2Cl^-(aq) \rightarrow 2H_2O(l) + Ba^{2+}(aq) + 2Cl^-(aq)$

34. The reaction $2K(s) + Br_2(l) \rightarrow 2KBr(s)$ is a(n) _____ reaction.
    a. precipitation
    b. acid-base
    c. oxidation-reduction
    d. double-displacement
    e. single-replacement

    ANSWER: c. oxidation-reduction

35. The reaction $AgNO_3(aq) + NaCl(aq) \rightarrow AgCl(s) + NaNO_3(aq)$ is a(n) _____ reaction.
    a. precipitation
    b. acid-base
    c. oxidation-reduction
    d. none of these
    e. single-replacement

    ANSWER: a. precipitation

36. The equation $2C_2H_6 + 7O_2 \rightarrow 4CO_2 + 6H_2O$ is an oxidation-reduction reaction. True or false?
    a. True, the carbon is oxidized and the oxygen is reduced.
    b. True, the carbon is reduced and the oxygen is oxidized.
    c. True, the carbon is oxidized and the hydrogen is reduced.
    d. True, the oxygen is reduced and the hydrogen is oxidized.
    e. False

    ANSWER: a.   True, the carbon is oxidized and the oxygen is reduced.

37. The reaction $HCl(aq) + KOH(aq) \rightarrow H_2O(l) + KCl(aq)$ is a(n) _____ reaction.
    a. precipitation
    b. acid-base
    c. oxidation-reduction
    d. single-replacement
    e. none of these

    ANSWER: b.   acid-base

38. How many electrons are transferred in the following oxidation-reduction reaction?
    $Zn(s) + 2AgNO_3(aq) \rightarrow Zn(NO_3)_2(aq) + 2Ag(s)$
    a. 1
    b. 2
    c. 3
    d. 4
    e. 5

    ANSWER: b.   2

39. How many of the following are oxidation-reduction reactions?
    I. reaction of a metal with a nonmetal
    II. synthesis
    III. combustion
    IV. precipitation
    V. decomposition
    a. 1
    b. 2
    c. 3
    d. 4
    e. 5

    ANSWER: d.   4

40. Electrons are transferred in combustion reactions. True or false?

    ANSWER: True

41. When a metal and a nonmetal react, the metal _____ electrons and the nonmetal _____ electrons.

    ANSWER: loses; gains

42. In what type of reaction is water always a product?
     a. precipitation
     b. acid-base
     c. oxidation
     d. decomposition
     e. synthesis

     ANSWER: b.   acid-base

43. Oxidation and reduction must occur simultaneously. True or false?

     ANSWER: True

44. Classify the following reaction:
     $2Mg(s) + O_2(g) \rightarrow 2MgO(s)$
     a. oxidation-reduction
     b. combustion
     c. synthesis
     d. two of the above
     e. a-c are all correct.

     ANSWER: e.   a-c are all correct.

45. Classify the following reaction:
     $HNO_3(aq) + KOH(aq) \rightarrow KNO_3(aq) + H_2O(l)$
     a. oxidation-reduction
     b. combustion
     c. precipitation
     d. acid-base
     e. two of the above

     ANSWER: d.   acid-base

46. When the following equation is balanced in standard form, what is the coefficient in front of the underlined substance?
     $C_2H_6(g) + O_2(g) \rightarrow \underline{CO_2(g)} + H_2O(l)$
     a. 1
     b. 2
     c. 3
     d. 4
     e. 5

     ANSWER: d.   4

47. A reaction that involves a transfer of electrons is called a(n) _____ reaction.
     a. precipitation
     b. acid-base
     c. oxidation-reduction
     d. double-displacement
     e. none of these

     ANSWER: c.   oxidation-reduction

48. Which of the following statements is *not* true?
  a. When a metal reacts with a nonmetal, an ionic compound is formed.
  b. A metal-nonmetal reaction can always be assumed to be an oxidation-reduction reaction.
  c. Two nonmetals can undergo an oxidation-reduction reaction.
  d. When two nonmetals react, the compound formed is ionic.
  e. A metal-nonmetal reaction involves electron transfer.

  ANSWER: d.   When two nonmetals react, the compound formed is ionic.

Use the following choices:
  a. oxidation-reduction
  b. acid-base
  c. precipitation
to classify each reaction given below (more than one choice may apply).

49. $ZnBr_2(aq) + 2AgNO_3(aq) \rightarrow Zn(NO_3)_2(aq) + 2AgBr(s)$

  ANSWER: c.   precipitation

50. $KBr(aq) + AgNO_3(aq) \rightarrow AgBr(s) + KNO_3(aq)$

  ANSWER: c.   precipitation

51. $HNO_3(aq) + NaOH(aq) \rightarrow H_2O(l) + NaNO_3(aq)$

  ANSWER: b.   acid-base

52. $H_2SO_4(aq) + Ba(OH)_2(aq) \rightarrow 2H_2O(l) + BaSO_4(s)$

  ANSWER: b.   acid-base;   c.   precipitation

53. $Zr(s) + O_2(g) \rightarrow ZrO_2(s)$

  ANSWER: a.   oxidation-reduction

54. $Zn(s) + 2HCl(aq) \rightarrow H_2(g) + ZnCl_2(aq)$

  ANSWER: a.   oxidation-reduction

55. $6Na(s) + N_2(g) \rightarrow 2Na_3N(s)$

  ANSWER: a.   oxidation-reduction

56. $Na_2SO_4(aq) + Pb(NO_3)_2(aq) \rightarrow PbSO_4(s) + 2NaNO_3(aq)$

  ANSWER: c.   precipitation

57. $2HCl(aq) + Pb(OH)_2(aq) \rightarrow PbCl_2(s) + 2H_2O(l)$

  ANSWER: b.   acid-base;   c.   precipitation

58. $2Hg(l) + O_2(g) \rightarrow 2HgO(s)$

ANSWER: a. oxidation-reduction

59. $HC_2H_3O_2(aq) + CsOH(aq) \rightarrow H_2O(l) + CsC_2H_3O_2(aq)$

ANSWER: b. acid-base

60. $Ca(s) + H_2(g) \rightarrow CaH_2(s)$

ANSWER: a. oxidation-reduction

61.   The equation $2Al(s) + 2Br_2(l) \rightarrow 2AlBr_3(s)$ is a(n) _____ reaction.
    a. oxidation-reduction and synthesis
    b. oxidation-reduction only
    c. synthesis only
    d. decomposition
    e. combustion

ANSWER: a.   oxidation-reduction and synthesis

62.   The equation $2Ag_2O(s) \rightarrow 4Ag(s) + O_2(g)$ is a(n) _____ reaction.
    a. oxidation-reduction
    b. synthesis
    c. decomposition
    d. combustion
    e. two of these

ANSWER: e.   two of these

63. $CH_4(g) + 2O_2(g) \rightarrow CO_2(g) + 2H_2O(g)$
    a. oxidation-reduction
    b. synthesis
    c. decomposition
    d. combustion
    e. two of these

ANSWER: e.   two of these

Use the following choices:
    a. oxidation-reduction
    b. combustion
    c. synthesis
    d. decomposition
to classify each of the following reactions (more than one choice may apply).

64. $2HgO(s) \rightarrow 2Hg(l) + O_2(g)$

ANSWER: a. oxidation-reduction;   d. decomposition

65. $2Na(s) + H_2(g) \rightarrow 2NaH(s)$

    ANSWER: a. oxidation-reduction;   c. synthesis

66. $C(s) + O_2(g) \rightarrow CO_2(g)$

    ANSWER: a. oxidation-reduction;   b. combustion;   c. synthesis

67. $2Cs(s) + F_2(g) \rightarrow 2CsF(s)$

    ANSWER: a. oxidation-reduction;   c. synthesis

68. $4NH_3(g) + 5O_2(g) \rightarrow 4NO(g) + 6H_2O(g)$

    ANSWER: a. oxidation-reduction;   b. combustion

69. $C_3H_8(g) + 5O_2(g) \rightarrow 3CO_2(g) + 4H_2O(g)$

    ANSWER: a. oxidation-reduction;   b. combustion

70. $2H_2O(l) \rightarrow 2H_2(g) + O_2(g)$

    ANSWER: a. oxidation-reduction;   d. decomposition

71. $S_8(s) + 12O_2(g) \rightarrow 8SO_3(g)$

    ANSWER: a. oxidation-reduction;   b. combustion;   c. synthesis

72. $P_4(s) + 5O_2(g) \rightarrow P_4O_{10}(s)$

    ANSWER: a. oxidation-reduction;   b. combustion;   c. synthesis

73. $2GaN(s) \rightarrow 2Ga(s) + N_2(g)$

    ANSWER: a. oxidation-reduction;   d. decomposition

CHAPTER 8
Chemical Composition

1. PVDF film is _____ and _____, properties that account for its many applications and uses.
    a. conductive; malleable
    b. piezoelectric; pyroelectric
    c. ductile; malleable
    d. conductive; ductile
    e. piezoelectric; isoelectric

    ANSWER: b. piezoelectric; pyroelectric

2. Because atoms are so _____, the standard units of mass are not useful in measurements.
    a. unstable
    b. scattered
    c. varied
    d. small
    e. electrically charged

    ANSWER: d. small

3. Choose the false statement.
    a. $1 \text{ mol} = 6.02 \times 10^{23}$ amu
    b. 6.02×10^{23} atoms = 1 mol of atoms
    c. 6.02×10^{23} hydrogen atoms weigh 1.008 g
    d. 1 mol of carbon atoms weighs 12.0 g
    e. Fluorine is a diatomic gas.

    ANSWER: a. $1 \text{ mol} = 6.02 \times 10^{23}$ amu

4. One mole of oxygen atoms represents
    a. 32.0 g
    b. 1.00 g
    c. 6.02×10^{23} atoms
    d. 16 atoms
    e. none of these

    ANSWER: c. 6.02×10^{23} atoms

5. Which represents the greatest number of atoms?
   a. 50.0 g Al
   b. 50.0 g Cu
   c. 50.0 g Zn
   d. 50.0 g Fe
   e. all the same

   ANSWER: a. 50.0 g Al

6. Which represents the greatest mass?
   a. 1.0 mol Al
   b. 1.0 mol Cu
   c. 1.0 mol Zn
   d. 1.0 mol Fe
   e. all the same

   ANSWER: c. 1.0 mol Zn

7. How many atoms of calcium are present in 80.0 g of calcium?
   a. 2
   b. 3.32×10^{-24}
   c. 6.02×10^{23}
   d. 1.20×10^{24}
   e. none of these

   ANSWER: d. 1.20×10^{24}

8. Calculate the molar mass of a sample if a single molecule weighs 5.34×10^{-23} g.
   a. 1.13×10^{46} g/mol
   b. 12.0 g/mol
   c. 5.34×10^{-23} g/mol
   d. 32.2 g/mol
   e. none of these

   ANSWER: d. 32.2 g/mol

9. A chemical mole
   a. contains 6.02×10^{23} particles
   b. is a large molecule
   c. contains an undetermined number of ions
   d. was a kind of material
   e. is no longer useful to chemistry

   ANSWER: a. contains 6.02×10^{23} particles

10. What is the mass of one atom of copper in grams?
    a. 63.5 g
    b. 52.0 g
    c. 58.9 g
    d. 65.4 g
    e. 1.06×10^{-22} g

    ANSWER: e. 1.06×10^{-22} g

11. Calculate the mass of 20.0 moles of He.
    a. 5.00
    b. 1.00
    c. 1.20×10^{25}
    d. 80.1
    e. 6.02×10^{23}

    ANSWER: d.   80.1

12. One atom of calcium weighs
    a. 20 amu
    b. 20 g
    c. 40.08 g
    d. 6.02×10^{23} amu
    e. none of these

    ANSWER: e.   none of these

13. How many atoms are there in 58.7 g of nickel?
    a. 1
    b. 28
    c. 6.02×10^{23}
    d. 1.204×10^{23}
    e. none of these

    ANSWER: c. 6.02×10^{23}

14. Which of the following contains the smallest number of moles?
    a. 1.0 g U
    b. 1.0 g P
    c. 1.0 g Al
    d. 1.0 g Br
    e. 1.0 g Na

    ANSWER: a.   1.0 g U

15. How many moles of Ca atoms are in 801 g Ca?
    a. 4.99×10^{-2} mol
    b. 20.0 mol
    c. 40.0 mol
    d. 1.60×10^3 mol
    e. 801 mol

    ANSWER: b.   20.0 mol

16. One mole of water weighs
    a. 3 g
    b. 18 g
    c. 1 g
    d. 10 g
    e. 18 mL

    ANSWER: b.   18 g

17. What is the mass of 2.00 moles of $Ca(OH)_2$?
    a. 74.1 g
    b. 56 g
    c. 122.5 g
    d. 222.4 g
    e. 148.2 g

    ANSWER: e.   148.2 g

18. A 20.0-g sample of Ca contains how many calcium atoms?
    a. 20.0 atoms
    b. 40.1 atoms
    c. 0.500 atoms
    d. 6.02×10^{23} atoms
    e. 3.0×10^{23} atoms

    ANSWER: e. 3.0×10^{23} atoms

19. A 1.0-mole sample of H_2O_2 weighs
    a. 1.0 g
    b. 34 g
    c. 17 g
    d. 18 g
    e. 18 amu

    ANSWER: b.   34 g

20. 0.314 mol of a diatomic molecule has a mass of 22.26 g. Identify the molecule.
    a. F_2
    b. Cl_2
    c. Br_2
    d. I_2
    e. none of these

    ANSWER: b. Cl_2

21. Calculate the mass of 8.43×10^{26} atoms of silver.

    ANSWER: 1.51×10^5 g

22. 12.4 g of Mg represents how many moles?

    ANSWER: 0.510 mol

23. 48.6 g of Cu represents how many moles?

    ANSWER: 0.765 mol

24. 3.82 g of Ca represents how many moles?

    ANSWER: 0.0953 mol

25. 58.7 g of Ni represents how many atoms?

    ANSWER: 6.02×10^{23} atoms

26. 84.2 g of Pt represents how many atoms?

    ANSWER: 2.60×10^{23} atoms

27. 10.4 g of Cr represents how many atoms?

    ANSWER: 1.20×10^{23} atoms

28. A sample containing 1.42 mol of zinc has a mass of _____ g.

    ANSWER: 92.8

29. A sample containing 14.3 mol of magnesium has a mass of _____ g.

    ANSWER: 348

30. A sample containing 0.241 mol of sodium has a mass of _____ g.

    ANSWER: 5.54

31. A 4.83 mol sample of aluminum represents how many atoms?

    ANSWER: 2.91×10^{24} atoms

32. A 6.52 mol sample of Zr represents how many atoms?

    ANSWER: 3.93×10^{24} atoms

33. A 11.4 mol sample of Co represents how many atoms?

    ANSWER: 6.87×10^{24} atoms

34. The molar mass of calcium phosphate is
    a. 324.99 g/mol
    b. 310.18 g/mol
    c. 175.13 g/mol
    d. 135.05 g/mol
    e. none of these

    ANSWER: b.   310.18 g/mol

35. What is the mass of 1.48 mol of potassium sulfide?
    a. 105 g
    b. 153 g
    c. 163 g
    d. 221 g
    e. none of these

    ANSWER: c.   163 g

36. Calculate the molar mass of barium nitrate.

    ANSWER: 261.3 g/mol

37. What is the molar mass of nitroglycerin, $C_3H_5(NO_3)_3$?
    a. 165 g/mol
    b. 227 g/mol
    c. 309 g/mol
    d. 199 g/mol
    e. none of these

    ANSWER: b.   227 g/mol

38. What is the molar mass of NaCl?
    a. 29.22 g/mol
    b. 175.3 g/mol
    c. 2.00 g/mol
    d. 58.44 g/mol
    e. 28 g/mol

    ANSWER: d.   58.44 g/mol

39. What is the molar mass of K_2SO_4?
     a. 86 g/mol
     b. 174.26 g/mol
     c. 87.13 g/mol
     d. 174×10^{23} g/mol
     e. 135.16 g/mol

     ANSWER: b. 174.26 g/mol

40. What is the molar mass of $Al(OH)_3$?
     a. 43.99 g/mol
     b. 88 g/mol
     c. 78.00 g/mol
     d. 156 g/mol
     e. 61 g/mol

     ANSWER: c. 78.00 g/mol

41. What is the molar mass of H_2S?
     a. 34.08 g/mol
     b. 78.16 g/mol
     c. 17 g/mol
     d. 51 g/mol
     e. 64.11 g/mol

     ANSWER: a. 34.08 g/mol

42. What is the molar mass of $C_{12}H_{22}O_{11}$?

     ANSWER: 342.30 g/mol

43. The molar mass of blood sugar, $C_6H_{12}O_6$, also known as glucose and dextrose, is
     a. 180 g/mol
     b. 6.02×10^{23} g/mol
     c. 29 g/mol
     d. 169 g/mol
     e. none of these

     ANSWER: a. 180 g/mol

44. 16.0 g of oxygen contains
     a. 1 mol of oxygen molecules
     b. 6.02×10^{23} oxygen atoms
     c. 32 amu
     d. 6.02×10^{23} oxygen molecules
     e. none of the above

     ANSWER: b. 6.02×10^{23} oxygen atoms

45. The molar mass of Na_3PO_4 is
    a. 8.0 g/mol
    b. 6×10^{23} g/mol
    c. 164 g/mol
    d. 118 g/mol
    e. 80 g/mol

    ANSWER: c. 164 g/mol

46. The mass in grams of 6.0 mol of hydrogen gas (contains H_2) is
    a. 6×10^{23} g
    b. 12 g
    c. 6.0 g
    d. 12×10^{23} g
    e. none of these

    ANSWER: b. 12 g

47. Which gas has the lowest mass per mole?
    a. nitrogen
    b. oxygen
    c. fluorine
    d. chlorine
    e. hydrogen

    ANSWER: e. hydrogen

48. The molar mass of magnesium hydroxide is
    a. 19.0 g/mol
    b. 30.0 g/mol
    c. 41.3 g/mol
    d. 42.3 g/mol
    e. 58.3 g/mol

    ANSWER: e. 58.3 g/mol

49. How many molecules of O_2 are there in 4.0 mol of O_2?
    a. 1.9×10^{23}
    b. 128
    c. 64
    d. 6.6×10^{-24}
    e. 2.4×10^{24}

    ANSWER: e. 2.4×10^{24}

50. Which of the following contains the smallest number of molecules?
    a. 5.0 g CO_2
    b. 5.0 g O_2
    c. 5.0 g N_2
    d. 5.0 g H_2O

    ANSWER: a.   5.0 g CO_2

51. The smallest unit of hydrogen fluoride gas is an example of a(n)
    a. mole
    b. molecule
    c. ion
    d. atom

    ANSWER: b.   molecule

52. The molar mass of $MgCl_2$ is
    a. 95.2 g/mol
    b. 59.8 g/mol
    c. 119.5 g/mol
    d. 125.9 g/mol
    e. none of these

    ANSWER: a.   95.2 g/mol

53. The mass of 0.80 mol of H_2 is
    a. 1.6 g
    b. 0.80 g
    c. 3.2 g
    d. 0.8 g
    e. none of these

    ANSWER: a.   1.6 g

54. 1.008 g of hydrogen gas contains
    a. 1 mol of hydrogen molecules, H_2
    b. 6.02×10^{23} hydrogen molecules, H_2
    c. 6.02×10^{23} hydrogen atoms
    d. 1 atom of hydrogen, H
    e. none of the above

    ANSWER: c. 6.02×10^{23} hydrogen atoms

55. The molar mass of calcium hydroxide is
    a. 57.6 g
    b. 97.6 g
    c. 74.1 g
    d. 56.1 g
    e. 29.8 g

    ANSWER: c.   74.1 g

56. One molecule of chlorine weighs
    a. 17 amu
    b. 34 amu
    c. 35.5 g
    d. 71 amu
    e. 71 g

    ANSWER: d. 71 amu

57. Calculate the mass, in grams, of 1.25×10^{26} formula units of sulfur dioxide.

    ANSWER: 1.33×10^4 g SO_2

58. Which gas has the lowest molar mass?
    a. N_2
    b. Cl_2
    c. O_2
    d. Br_2
    e. Ne

    ANSWER: e. Ne

59. A mole of sulfur trioxide molecules contains how many oxygen atoms?
    a. 32
    b. 3.01×10^{23}
    c. 6.02×10^{23}
    d. 1.204×10^{24}
    e. 1.807×10^{24}

    ANSWER: e. 1.807×10^{24}

60. A mole of oxygen molecules contains how many oxygen atoms?
    a. 16
    b. 32
    c. 3.01×10^{23}
    d. 6.02×10^{23}
    e. 1.204×10^{24}

    ANSWER: e. 1.204×10^{24}

61. Calculate the mass of 9.13×10^{19} molecules of HCl.

    ANSWER: 5.53×10^{-3} g

62. The number of moles in 12.0 g of C_2H_6O (molar mass = 46.1 g/mol) is calculated as follows:
    a. 12.0/24.0
    b. 12.0 x 24.0
    c. 12.0/46.1
    d. 12.0 x 46.1

    ANSWER: c.   12.0/46.1

63. Calculate the number of moles of water molecules of 25.0 g of water.
    a. 4.17 mol
    b. 6.98 mol
    c. 0.720 mol
    d. 1.39 mol
    e. 2.78 mol

    ANSWER: d.   1.39 mol

64. 6.00 g of water contains how many moles of water?
    a. 108 mol
    b. 18.0 mol
    c. 3.00 mol
    d. 6.00 mol
    e. 0.333 mol

    ANSWER: e.   0.333 mol

65. 6.00 g of water contains how many molecules of water?
    a. 2.10×10^{23}
    b. 4.20×10^{23}
    c. 0.333
    d. 3.00
    e. 3.61×10^{24}

    ANSWER: a. 2.10×10^{23}

66. Calculate the mass of 3.50 mol of sulfur dioxide.
    a. 112 g
    b. 224 g
    c. 61 g
    d. 2.10×10^{24} g
    e. 4.1×10^{24} g

    ANSWER: b.   224 g

67. Convert 48 g O_2 to mol O_2.
    a. 1.5 mol
    b. 3.0 mol
    c. 0.75 mol
    d. 3×10^{23} mol
    e. 1.5×10^{-23} mol

    ANSWER: a.   1.5 mol

68. Convert: 7.14 mol K_2O = _____ g K_2O

    ANSWER: 673

69. Convert: 0.189 mol $Ca(OH)_2$ = _____ g $Ca(OH)_2$

    ANSWER: 14.0

70. Convert: 4.89 g O_3 = _____ molecules O_3

    ANSWER: 6.13×10^{22}

71. Convert: 6.12 mol $Cu(NO_3)_2$ = _____ g $Cu(NO_3)_2$

    ANSWER: 1150

72. Convert: 42.3 g NO_2 = _____ moles NO_2

    ANSWER: 0.919

73. Convert: 9.14 mol PCl_5 = _____ molecules PCl_5

    ANSWER: 5.50×10^{24}

74. Convert: 72.8 g CO = _____ molecules CO

    ANSWER: 1.57×10^{24}

75. Convert: 1.42 mol $Mg(NO_3)_2$ = _____ g $Mg(NO_3)_2$

    ANSWER: 211

76. Convert: 12.4 mol CaF_2 = _____ g CaF_2

    ANSWER: 968

77. Convert: 14.3 g NaCl = _____ mol NaCl

    ANSWER: 0.245

78. Convert: 6.18×10^{22} molecules NH_3 = _____ mol NH_3

    ANSWER: 0.103

79. Convert: 243 g $MgCl_2$ = _____ units of $MgCl_2$

    ANSWER: 1.54×10^{24}

80. A 98.6 -g sample of SO_2 contains how many moles of SO_2?

    ANSWER: 1.54 mol

81. Calculate the mass of 1.42 mol of silver nitrate.

    ANSWER: 241 g

82. A 50.0-g sample of H_2O contains how many molecules of water?

    ANSWER: 1.67×10^{24} molecules

83. A 75.0-mL sample of Hg (density = 13.6 g/mL) contains how many atoms of Hg?

    ANSWER: 3.06×10^{24} atoms

84. A sample with 1.25×10^{24} atoms of copper represents how many moles of copper?

    ANSWER: 2.08 mol

85. How many molecules are present in 3.91×10^{-4} g of H_2?

    ANSWER: 1.17×10^{20} molecules

86. How many moles of Al_2O_3 are present in 93.6 g of this compound?

    ANSWER: 0.918 mol

87. Calculate the number of moles in 3.19 g of $Rb_2C_2O_4$.

    ANSWER: 0.0123 mol

88. Calculate the number of molecules in 0.000108 g of gaseous oxygen.

    ANSWER: 2.03×10^{18} molecules

89. Calculate the number of moles of CH_4 in 84 g CH_4.

    ANSWER: 5.3 mol

90. Calculate the number of molecules of CH_4 in 48 g CH_4.

    ANSWER: 1.8×10^{24} molecules

91. Calculate the mass of 0.500 mol of H_2SO_4.

    ANSWER: 49.0 g

92. Calculate the mass of sulfur in 2.0 mol of H_2SO_4.

ANSWER: 64 g

93. Calculate the number of moles of Cl_2 molecules in a sample that contains 7.12×10^{25} molecules of Cl_2.

ANSWER: 118 mol

94. Compound X_2Y is known to be 60% X by mass. Calculate the percent Y by mass of the compound X_2Y_2.
   a. 20%
   b. 30%
   c. 40%
   d. 60%
   e. 80%

ANSWER: d. 60%

95. Arrange the following compounds in order of increasing percentage of oxygen by mass.
   a. NO_2, CO_2, $C_2H_4O_2$, O_2, H_2O
   b. $C_2H_4O_2$, O_2, NO_2, CO_2, H_2O
   c. $C_2H_4O_2$, NO_2, CO_2, O_2, H_2O
   d. NO_2, O_2, H_2O, CO_2, $C_2H_4O_2$
   e. $C_2H_4O_2$, NO_2, CO_2, H_2O, O_2

ANSWER: e. $C_2H_4O_2$, NO_2, CO_2, H_2O, O_2

96. What is the empirical formula of an oxide of bromine that contains 71.4% bromine?
   a. BrO_3
   b. BrO_2
   c. BrO
   d. Br_2O
   e. none of these

ANSWER: b. BrO_2

97. Consider separate 100.0-g samples of each of the following: NH_3, N_2O, HCN, N_2H_4, and HNO_3. Which of the samples has the greatest mass of nitrogen?
   a. NH_3
   b. N_2O
   c. HCN
   d. N_2H_4
   e. HNO_3

ANSWER: d. N_2H_4

98. Consider separate 100.0-g samples of each of the following: NH_3, N_2O, HCN, N_2H_4, and HNO_3. Which of the samples has the least mass of nitrogen?
    a. NH_3
    b. N_2O
    c. HCN
    d. N_2H_4
    e. HNO_3

    ANSWER: e. HNO_3

99. A hydrocarbon has the formula C_2H_4. What is the percent by mass of carbon in the compound?
    a. 33.3%
    b. 50.0%
    c. 62.4%
    d. 85.6%
    e. 92.1%

    ANSWER: d.   85.6%

100. Calculate the percentage composition (by mass) of all the elements in $Cd_3(AsO_4)_2$.

    ANSWER: 54.8% Cd; 24.4% As; 20.8% O

101. Determine the percentage composition (by mass) of H_2SO_4.

    ANSWER: 2.06% H; 32.69% S; 65.25% O

102. The mass percent of oxygen in CaO is
    a. 28.5%
    b. 50%
    c. 72.4%
    d. 25.0%
    e. Cannot be determined from the information given

    ANSWER: a.   28.5%

103. Determine the percentage composition (by mass) of NH_4NO_3.

    ANSWER: 35.00% N; 59.96% O; 5.04% H

104. Determine the percentage composition (by mass) of tin in tin(II) fluoride, to three significant figures.

    ANSWER: 75.7% Sn

105. Which of the following has the largest percent by mass of carbon?
    a. $CaCO_3$
    b. CO_2
    c. CH_4
    d. $NaHCO_3$

    ANSWER: c. CH_4

106. The mass percent of nitrogen in NH_4Cl is
    a. 16.7%
    b. 25.0%
    c. 26.2%
    d. none of these

    ANSWER: c.   26.2%

107. Which of the following has the highest mass percentage of nitrogen?
    a. NH_3
    b. N_2H_4
    c. HNO_3
    d. NH_4NO_3
    e. NO

    ANSWER: b. N_2H_4

108. What is the percent (by mass) of carbon in glucose, $C_6H_{12}O_6$?
    a. 40.0%
    b. 20.0%
    c. 80.0%
    d. 6.70%
    e. none of these

    ANSWER: a.   40.0%

109. Which of the following has the highest mass percent of hydrogen?
    a. CH_4
    b. SiH_4
    c. GeH_4
    d. SnH_4
    e. All have the same mass percent of hydrogen.

    ANSWER: a. CH_4

110.   Which of the following compounds contains the lowest mass percent of oxygen?
a.   SnO
b. SnO_2
c.   CaO
d.   MgO
e. H_2O_2

ANSWER: a.   SnO

111.   A sample of metal weighing 2.51 g is combined with oxygen; the metal oxide weighs 3.01 g. The mass percent of oxygen in the compound is
a.   19.9%
b.   16.6%
c.   83.4%
d.   50.0%
e.   cannot be determined from information given

ANSWER: b.   16.6%

112.   Which of the following has the empirical formula CH_2?
a. C_2H_6
b. H_2CO_3
c. C_6H_6
d. C_6H_{12}
e. C_2H_4O

ANSWER: d. C_6H_{12}

113.   Choose the pair of compounds with the same empirical formula.
a. C_2H_2 and C_6H_6
b. $NaHCO_3$ and Na_2CO_3
c. K_2CrO_4 and $K_2Cr_2O_7$
d. H_2O and H_2O_2

ANSWER: a. C_2H_2 and C_6H_6

114.   The empirical formula for the compound having the formula $H_2C_2O_4$ is
a.   COH
b. COH_2
c. C_2H_2
d. $C_2O_4H_2$
e. CO_2H

ANSWER: e. CO_2H

115. Calculate the empirical formula of a compound that is 85.6% C and 14.4% H (by mass).
    a. C_2H_4
    b. C_2H
    c. CH_2
    d. CH
    e. C_3H_5

    ANSWER: c. CH_2

116. A compound is analyzed and found to contain 12.1% carbon, 16.2% oxygen, and 71.7% chlorine (by mass). Calculate the empirical formula of this compound.
    a. $COCl$
    b. $COCl_2$
    c. CO_2Cl
    d. CO_2Cl_2
    e. $COCl_4$

    ANSWER: b. $COCl_2$

117. A compound contains 40.0% carbon, 6.7% hydrogen, and 53.3% oxygen (by mass). Calculate the empirical formula.
    a. CH_2O
    b. C_2H_2O
    c. CH_4O
    d. $C_3H_6O_3$
    e. C_2HO_2

    ANSWER: a. CH_2O

118. A compound contains 25.94% N and 74.06% O (by mass). What is the empirical formula?

    ANSWER: N_2O_5

119. Calculate the empirical formula of a compound containing 18.29% H and 81.71% C (by mass).

    ANSWER: C_3H_8

120. Determine the empirical formula of a compound containing 54.2% F and 45.8% S (by mass).

    ANSWER: SF_2

121. A compound has 40.68% carbon, 5.12% hydrogen, and 54.20% oxygen (by mass). Calculate its empirical formula.

    ANSWER: $C_2H_3O_2$

122. Calculate the empirical formula of a compound that is 50.04% C, 5.59% H, and 44.37% O (by mass).

    ANSWER: $C_3H_4O_2$

123. A 7.33-g sample of lanthanum, La, combines with oxygen to give 10.29 g of the oxide. Calculate the empirical formula of this oxide.

ANSWER: La_2O_7

124. Calculate the molecular formula of a compound with the empirical formula CH_2O and a molar mass of 150 g/mol.
   a. $C_2H_4O_2$
   b. $C_3H_6O_3$
   c. $C_4H_8O_4$
   d. $C_5H_{10}O_5$
   e. $C_6H_{12}O_6$

ANSWER: d. $C_5H_{10}O_5$

125. Acetylene gas is 92.3% carbon and 7.7% hydrogen (by mass), and its molar mass is 26 g/mol. What is its molecular formula?
   a. CH
   b. C_2H_2
   c. CH_4
   d. C_4H_4
   e. none of these

ANSWER: b. C_2H_2

126. The empirical formula of a compound is known to be CH_2, and its molar mass is 56 g/mol. What is the molecular formula?

ANSWER: C_4H_8

127. The empirical formula of a compound is CH_2O, and its mass is 120 amu/molecule. Calculate its molecular formula.
   a. CH_2O
   b. $C_2H_4O_2$
   c. $C_3H_6O_3$
   d. $C_4H_8O_4$
   e. none of these

ANSWER: d. $C_4H_8O_4$

128. A compound contains 12.8% C, 2.1% H, and 85.1% Br (by mass). Calculate the empirical formula and the molecular formula of this compound given that the molar mass is 188 g/mol.

ANSWER:
CH_2Br
$C_2H_4Br_2$

129. A compound contains 10.13% C and 89.87% Cl (by mass). Determine both the empirical formula and the molecular formula of the compound given that the molar mass is 237 g/mol.

ANSWER:
CCl_3
C_2Cl_6

130. A certain compound has an empirical formula of NH_2O. Its molar mass is between 55 and 65 g/mol. Its molecular formula is
   a. NH_2O
   b. $N_2H_2O_2$
   c. $N_2H_4O_2$
   d. not calculable

ANSWER: c. $N_2H_4O_2$

131. A compound has a molar mass of 86 g/mol and has the percent composition (by mass) of 55.8% C, 37.2% O, and 7.0% H. Determine the empirical formula and the molecular formula.

ANSWER:
C_2H_3O
$C_4H_6O_2$

132. A compound has a molar mass of 100 g/mol and the percent composition (by mass) of 65.45% C, 5.45% H, and 29.09% O. Determine the empirical formula and the molecular formula.
   a. CHO and $C_6H_6O_6$
   b. C_3H_3O and $C_6H_6O_2$
   c. C_3HO and $C_6H_2O_2$
   d. CH_2O and $C_4H_8O_4$
   e. CH_4O and $C_3H_{12}O_3$

ANSWER: b. C_3H_3O and $C_6H_6O_2$

133. The empirical formula for acetic acid is CH_2O. Its molar mass is 60 g/mol. The molecular formula is
   a. CH_2O
   b. $C_2H_4O_2$
   c. C_2H_6O
   d. C_2HO_2
   e. none of the above

ANSWER: b. $C_2H_4O_2$

134. A certain compound is found to have the percent composition (by mass) of 85.63% C and 14.37% H. The molar mass of the compound was found to be 42.0 g/mol. Calculate the empirical and the molecular formulas.

a. C_2H_3 and C_4H_6

b. CH and C_3H_3

c. CH_2 and C_3H_6

d. CH_3 and C_2H_6

e. C_2H_6 and C_3H_9

ANSWER: c. CH_2 and C_3H_6

CHAPTER 9
Chemical Quantities

1. The balanced equation $P_4(s) + 6H_2(g) \rightarrow 4PH_3(g)$ tells us that 2 mol H_2
    a. reacts with 1 mol P_4
    b. produces 4 mol PH_3
    c. cannot react with phosphorus
    d. produces 2 mol PH_3
    e. reacts with 2 mol P_4

    ANSWER: d. produces 2 mol PH_3

2. The equation $N_2(g) + 3H_2(g) \rightarrow 2NH_3(g)$ can be interpreted by saying that 1 mol of N_2 reacts with 3 mol of H_2 to form 2 mol of NH_3. T/F _____

    ANSWER: True

3. A balanced chemical equation is one that has the same number of moles of molecules on each side of the equation. T/F _____

    ANSWER: False

4. The balanced equation $2Cu(s) + O_2(g) \rightarrow 2CuO(s)$ tells us that 1 mol of Cu
    a. reacts with 1 mol of O_2
    b. produces 1 mol of CuO
    c. must react with 32 g of O_2
    d. cannot react with oxygen
    e. produces 2 mol of CuO

    ANSWER: b. produces 1 mol of CuO

5. A 3.0-mol sample of $KClO_3$ was decomposed according to the equation
    $2KClO_3(s) \rightarrow 2KCl(s) + 3O_2(g)$
    How many moles of O_2 are formed assuming 100% yield?
    a. 2.0 mol
    b. 2.5 mol
    c. 3.0 mol
    d. 4.0 mol
    e. 4.5 mol

    ANSWER: e. 4.5 mol

6. An excess of Al and 6.0 mol of Br_2 are reacted according to the equation
   $2Al + 3Br_2 \rightarrow 2AlBr_3$
   How many moles of $AlBr_3$ will be formed assuming 100% yield?
   a. 2.0 mol
   b. 3.0 mol
   c. 4.0 mol
   d. 6.0 mol
   e. 8.0 mol

   ANSWER: c.   4.0 mol

7. The rusting of iron is represented by the equation $4Fe + 3O_2 \rightarrow 2Fe_2O_3$. If you have a 1.50-mol sample of iron, how many moles of Fe_2O_3 will there be after the iron has rusted completely?
   a. 0.50 mol
   b. 0.75 mol
   c. 1.0 mol
   d. 1.50 mol
   e. 2.0 mol

   ANSWER: b.   0.75 mol

8. For the reaction
   $C_2H_4(g) + 3O_2(g) \rightarrow 2CO_2(g) + 2H_2O(g)$
   if 6.0 mol of CO_2 are produced, how many moles of O_2 were reacted?
   a. 4.0 mol
   b. 7.5 mol
   c. 9.0 mol
   d. 15.0 mol
   e. none of these

   ANSWER: c.   9.0 mol

9. A mole ratio is used to convert the moles of starting substance to the moles of desired substance. T/F _____

   ANSWER: True

10. The equation for a reaction should be balanced before doing stoichiometric calculations. T/F _____

    ANSWER: True

Refer to the following equation: $4NH_3(g) + 7O_2(g) \rightarrow 4NO_2(g) + 6H_2O(g)$

11. How many moles of ammonia will be required to produce 10.0 mol of water?
    a. 4.00 mol
    b. 10.0 mol
    c. 6.67 mol
    d. 5.00 mol
    e. none of these

    ANSWER: c.   6.67 mol

12. How many molecules of NO_2 are produced when 1 mol of ammonia is completely reacted?
    a. 4
    b. 12.044×10^{23}
    c. 6.022×10^{23}
    d. 46
    e. none of these

    ANSWER: c. 6.022×10^{23}

13. How many molecules of water are produced for each mole of NO_2 given off?
    a. 12.044×10^{23}
    b. 6.022×10^{23}
    c. 18
    d. 9.033×10^{23}
    e. none of these

    ANSWER: d. 9.033×10^{23}

14. In the reaction $N_2(g) + 3H_2(g) \rightarrow 2NH_3(g)$, how many moles of ammonia would be produced from 1.0 mol of hydrogen and excess nitrogen?
    a. 1.3 mol
    b. 3.0 mol
    c. 0.67 mol
    d. 2.0 mol
    e. 0.33 mol

    ANSWER: c.   0.67 mol

Refer to the following unbalanced equation:
$$C_6H_{14} + O_2 \rightarrow CO_2 + H_2O$$

15. When balanced in standard form (smallest whole numbers), the coefficient for CO_2 is
    a. 6
    b. 8
    c. 10
    d. 12
    e. 14

    ANSWER: d.   12

16. What mass of oxygen (O_2) is required to react completely with 25.0 g of C_6H_{14}?
    a. 9.28 g
    b. 16.0 g
    c. 32.0 g
    d. 88.2 g
    e. 608 g

    ANSWER: d.   88.2 g

17. What mass of carbon dioxide (CO_2) can be produced from 25.0 g of C_6H_{14} and excess oxygen?
    a. 12.8 g
    b. 44.0 g
    c. 76.6 g
    d. 264 g
    e. 528 g

    ANSWER: c.   76.6 g

18. How many molecules of carbon dioxide would be formed if 6.75 g of propane is burned in the following reaction?
    $C_3H_8(g) + 5O_2(g) \rightarrow 3CO_2(g) + 4H_2O(g)$
    a. 5.54×10^{23} molecules
    b. 1.39×10^{23} molecules
    c. 20.3×10^{23} molecules
    d. 2.77×10^{23} molecules
    e. 3.89×10^{23} molecules

    ANSWER: d. 2.77×10^{23} molecules

19. What mass of carbon dioxide would be produced when 10.0 g of butane reacts with an excess of oxygen in the following reaction?
    $2C_4H_{10}(g) + 13O_2(g) \rightarrow 8CO_2(g) + 10H_2O(g)$
    a. 7.57 g CO_2
    b. 30.3 g CO_2
    c. 40.0 g CO_2
    d. 352 g CO_2
    e. none of these

    ANSWER: b.   30.3 g CO_2

20. Calculate the mass of water produced when 5.00 g of methane, CH_4, reacts with an excess of oxygen in the following unbalanced reaction.
    $CH_4(g) + O_2(g) \rightarrow CO_2(g) + H_2O(g)$
    a. 5.62 g H_2O
    b. 10.0 g H_2O
    c. 11.2 g H_2O
    d. 18.0 g H_2O
    e. 36.0 g H_2O

    ANSWER: c.   11.2 g H_2O

21. Calculate the mass of carbon dioxide produced from 11.2 g of octane, C_8H_{18}, in the following reaction.
    $$2C_8H_{18}(g) + 25O_2(g) \rightarrow 16CO_2(g) + 18H_2O(g)$$
    a. 34.5 g CO_2
    b. 89.6 g CO_2
    c. 17.3 g CO_2
    d. 46.2 g CO_2
    e. 58.9 g CO_2

    ANSWER: a.   34.5 g CO_2

22. Calculate the molecules of oxygen required to react with 16.0 g of sulfur in the following reaction.
    $$2S(s) + 3O_2(g) \rightarrow 2SO_3(g)$$

    ANSWER: 4.51×10^{23} molecules O_2

23. Consider the reaction
    $$2Fe(s) + 3O_2(g) \rightarrow Fe_2O_3(s)$$
    If 12.5 g of iron(III) oxide (rust) is produced from 8.74 g of iron, how many grams of oxygen are needed for this reaction?
    a. 12.5 g
    b. 7.5 g
    c. 8.74 g
    d. 21.2 g
    e. none of these

    ANSWER: b.   7.5 g

24. In the reaction
    $$3H_2(g) + N_2(g) \rightarrow 2NH_3(g)$$
    how many molecules of hydrogen are required to react with 7.00 g of nitrogen?

    ANSWER: 4.52×10^{23} molecules H_2

25. How many moles of oxygen are produced by decomposing 41.1 g of H_2O_2 (molar mass = 34.0 g/mol) according to the equation
    $$2H_2O_2(l) \rightarrow 2H_2O(l) + O_2(g)$$

    ANSWER: 0.604 mol O_2

26. In the reaction
    $$Cu(s) + 2AgNO_3(aq) \rightarrow 2Ag(s) + Cu(NO_3)_2(aq)$$
    what number of grams of silver can be produced from 49.1 g of copper?

    ANSWER: 167 g Ag

27. How many atoms of aluminum can be produced by the decomposition of 33.3 g of Al_2O_3? (*Hint:* Write and balance the equation first.)

    ANSWER: 3.93×10^{23} Al atoms

28. Nitrogen and hydrogen gases are combined at high temperatures and pressures to produce ammonia, NH_3. If 100. g of N_2 is reacted with excess H_2, what number of moles of NH_3 will be formed?
    a. 2.00 mol
    b. 3.57 mol
    c. 7.14 mol
    d. 33.3 mol
    e. none of these

    ANSWER: c.   7.14 mol

29. Methane, CH_4, the major component of natural gas burns in air to form CO_2 and H_2O. What mass of water is formed in the complete combustion of 5.00×10^3 g of CH_4?
    a. 5.00×10^3 g
    b. 1.00×10^4 g
    c. 5.65×10^3 g
    d. 1.12×10^4 g
    e. none of these

    ANSWER: d. 1.12×10^4 g

30. How many moles of O_2 are required for the complete reaction of 45 g of C_2H_4 to form CO_2 and H_2O?
    a. 1.3×10^2 mol
    b. 0.64 mol
    c. 112.5 mol
    d. 4.8 mol
    e. none of these

    ANSWER: d.   4.8 mol

31. If 18.0 g of CO_2 is produced in the reaction of C_2H_2 with O_2 to form CO_2 and H_2O, how many grams of H_2O are produced in this reaction?
    a. 7.37 g
    b. 3.68 g
    c. 9.0 g
    d. 14.7 g
    e. none of these

    ANSWER: b.   3.68 g

32. What number of moles of ammonia can be produced from 8.0 g of hydrogen gas and excess nitrogen gas?

    ANSWER: 2.65 mol NH_3

33. When 1.0 mol of Fe reacts with Cl_2 according to the equation

$$2Fe + 3Cl_2 \rightarrow 2FeCl_3$$

how many moles of Cl_2 are required to react with all the iron?

a. 1.0 mol
b. 0.67 mol
c. 1.5 mol
d. 3.0 mol
e. none of these

ANSWER: c.   1.5 mol

34. In the reaction

$$3Cl_2(g) + 6NaOH(aq) \rightarrow 5NaCl(aq) + NaClO_3(aq) + 3H_2O(l)$$

how many molecules of water can be produced starting with 23 kilograms of sodium hydroxide and excess Cl_2?

ANSWER: 1.7×10^{26} molecules H_2O

35. In the reaction

$$3Cl_2(g) + 6NaOH(aq) \rightarrow 5NaCl(aq) + NaClO_3(aq) + 3H_2O(l)$$

how many moles of chlorine molecules are needed to react with 10.9 g of NaOH?

ANSWER: 0.136 mol Cl_2

36. In the reaction

$$3Cl_2(g) + 6NaOH(aq) \rightarrow 5NaCl(aq) + NaClO_3(aq) + 3H_2O(l)$$

how many grams of sodium chloride can be produced from 10.9 g of NaOH?

ANSWER: 13.3 g NaCl

37. For the reaction

$$2S(s) + 3O_2(g) \rightarrow 2SO_3(g)$$

how many moles of SO_3 will be produced from 2.0 mol O_2 and excess S?

ANSWER: 1.3 mol SO_3

38. For the reaction

$$2S(s) + 3O_2(g) \rightarrow 2SO_3(g)$$

how many moles of SO_3 can be produced from 8.0 g O_2 and excess S?

ANSWER: 0.17 mol SO_3

39. For the reaction

$$CaCO_3(s) + 2HCl(aq) \rightarrow CaCl_2(aq) + CO_2(g) + H_2O(l)$$

how many grams of $CaCl_2$ could be obtained if 15.0 g HCl is allowed to react with excess $CaCO_3$?

ANSWER: 22.8 g $CaCl_2$

40. For the reaction
    $$2Cl_2(g) + 4NaOH(aq) \rightarrow 3NaCl(aq) + NaClO_2(aq) + 2H_2O(l)$$
    how many moles of Cl_2 are needed to react with 14.4 g NaOH?

    ANSWER: 0.180 mol Cl_2

41. For the reaction
    $$2Cl_2(g) + 4NaOH(aq) \rightarrow 3NaCl(aq) + NaClO_2(aq) + 2H_2O(l)$$
    how many grams of NaCl can be produced from 10.9 g of Cl_2 and excess NaOH?

    ANSWER: 13.5 g NaCl

42. For the reaction
    $$2Cl_2(g) + 4NaOH(aq) \rightarrow 3NaCl(aq) + NaClO_2(aq) + 2H_2O(l)$$
    how many molecules of H_2O can be produced from 23 g of NaOH and excess Cl_2?

    ANSWER: 1.7×10^{23} molecules H_2O

43. Consider the reaction
    $$Mg_2Si(s) + 2H_2O(l) \rightarrow 2Mg(OH)_2(aq) + SiH_4(g)$$
    Calculate the number of grams of silane gas, SiH_4, formed if 25.0 g of Mg_2Si reacts with excess H_2O.

    ANSWER: 10.5 g SiH_4

44. Fe_3O_4 reacts with CO according to the reaction
    $$Fe_3O_4(s) + 4CO(g) \rightarrow 4CO_2(g) + 3Fe(s)$$
    If 201 g Fe_3O_4 is reacted with excess CO, what mass of CO_2 will be produced?

    ANSWER: 153 g CO_2

45. Fe_3O_4 reacts with CO according to the reaction
    $$Fe_3O_4(s) + 4CO(g) \rightarrow 4CO_2(g) + 3Fe(s)$$
    If 234 g CO is reacted with excess Fe_3O_4, what mass of CO_2 will be produced?

    ANSWER: 368 g CO_2

46. Consider the reaction
    $$2CH_4(g) + 3O_2(g) + 2NH_3(g) \rightarrow 2HCN(g) + 6H_2O(g)$$
    If 128 g NH_3 is reacted with excess CH_4 and O_2, what mass of HCN can be produced?

    ANSWER: 203 g HCN

47. Fe_2O_3 (molar mass = 159.7 g/mol) reacts with CO (molar mass = 28.0 g/mol) according to the equation

    $Fe_2O_3(s) + 3CO(g) \rightarrow 3CO_2(g) + 2Fe(s)$

    When 352 g Fe_2O_3 reacts with excess CO, what number of moles of Fe (iron) is produced?

    ANSWER: 4.41 mol Fe

48. Fe_2O_3 (molar mass = 159.7 g/mol) reacts with CO (molar mass = 28.0 g/mol) according to the equation

    $Fe_2O_3(s) + 3CO(g) \rightarrow 3CO_2(g) + 2Fe(s)$

    When 116 g of CO reacts with excess Fe_2O_3, how many moles of iron (Fe) will be produced?

    ANSWER: 2.76 mol Fe

49. For the reaction

    $2S(s) + 3O_2(g) \rightarrow 2SO_3(g)$

    if 6.3 g of S is reacted with 10.0 g of O_2, show by calculation which one will be the limiting reactant.

    ANSWER: S is the limiting reactant.

50. For the reaction

    $CaCO_3(s) + 2HCl(aq) \rightarrow CaCl_2(aq) + CO_2(g) + H_2O(l)$

    68.1 g solid $CaCO_3$ is mixed with 51.6 g HCl. What number of grams of CO_2 will be produced?
    a. 15.0 g CO_2
    b. 59.8 g CO_2
    c. 33.7 g CO_2
    d. 69.4 g CO_2
    e. 29.9 g CO_2

    ANSWER: a.   15.0 g CO_2

51. For the reaction

    $2Cl_2(g) + 4NaOH(aq) \rightarrow 3NaCl(aq) + NaClO_2(aq) + 2H_2O(l)$

    11.9 g Cl_2 is reacted with 12.0 g NaOH. Determine which is the limiting reactant.

    ANSWER: NaOH is the limiting reactant.

52. Consider the reaction

    $Mg_2Si(s) + 4H_2O(l) \rightarrow 2Mg(OH)_2(aq) + SiH_4(g)$

    Which of the reactants is in excess if we start with 50.0 g of each reactant?

    ANSWER: H_2O is in excess.

53. Sodium and water react according to the reaction

    $2Na(s) + 2H_2O(l) \rightarrow 2NaOH(aq) + H_2(g)$

    What number of moles of H_2 will be produced when 4 mol Na is added to 2 mol H_2O?

    a. 1 mol
    b. 2 mol
    c. 3 mol
    d. 4 mol
    e. none of these

    ANSWER: a.   1 mol

54. For the reaction of $C_2H_4(g)$ with $O_2(g)$ to form $CO_2(g)$ and $H_2O(g)$, what number of moles of CO_2 can be produced by the reaction of 5.00 mol C_2H_4 and 12.0 mol O_2?

    a. 4.00 mol
    b. 5.00 mol
    c. 8.00 mol
    d. 10.0 mol
    e. none of these

    ANSWER: c.   8.00 mol

55. For the reaction of $C_2H_4(g)$ with $O_2(g)$ to form $CO_2(g)$ and $H_2O(g)$, what number of moles of CO_2 can be produced by the reaction of 0.480 mol of C_2H_4 and 1.08 mol of O_2?

    a. 0.240 mol
    b. 0.960 mol
    c. 0.720 mol
    d. 0.864 mol
    e. none of these

    ANSWER: c.   0.720 mol

56. For the reaction of $C_2H_4(g)$ with $O_2(g)$ to form $CO_2(g)$ and $H_2O(g)$, what number of grams of CO_2 could be produced from 2.0 g of C_2H_4 and 5.0 g of O_2?

    a. 5.5 g
    b. 4.6 g
    c. 7.6 g
    d. 6.3 g
    e. none of these

    ANSWER: b.   4.6 g

57. Determine the limiting reactant when 4.00 mol of Sb are reacted with 5.00 mol Cl according to the unbalanced equation

    $Sb + Cl_2 \rightarrow SbCl_3$

    a. Sb
    b. Cl
    c. Cl_2
    d. $SbCl_3$
    e. no limiting reactant

    ANSWER: c. Cl_2

58. Determine the limiting reactant when 50.0 g of CaO are reacted with 50.0 g of C according to the unbalanced equation
    $CaO + C \rightarrow CaC_2 + CO_2$
    a. CaO
    b. C
    c. CaC_2
    d. CO_2
    e. no limiting reactant

    ANSWER: a. CaO

59. Consider the equation
    $2A + 3B \rightarrow C$
    If 4.0 mol of A is reacted with 4.0 mol of B, which of the reactants is limiting?
    a. Neither is limiting because equal amounts (4.0 mol) of each reactant are reacted.
    b. A is limiting because 2 is smaller than 3 (the numbers refer to the coefficients in the balanced equation).
    c. A is limiting because there are 2 mol and 4.0 mol are needed.
    d. B is limiting because 3 is larger than 2 (the numbers refer to the coefficients in the balanced equation).
    e. B is limiting because there are 4.0 mol and 6.0 mol are needed.

    ANSWER: e. B is limiting because there are 4.0 mol and 6.0 mol are needed.

60. The limiting reactant in a reaction is
    a. the reactant for which there is the least amount in grams
    b. the reactant which has the lowest coefficient in a balanced equation
    c. the reactant for which there is the most amount in grams
    d. the reactant for which there is the fewest number of moles
    e. none of the above

    ANSWER: e. none of the above

61. The limiting reactant is the reactant
    a. for which you have the lowest mass in grams.
    b. that has the lowest coefficient in the balanced equation.
    c. that has the lowest molar mass.
    d. that is left over after the reaction has gone to completion.
    e. none of the above

    ANSWER: e. none of the above

For questions 62-64, consider that calcium metal reacts with oxygen gas in the air to form calcium oxide. Suppose we react 6.00 mol calcium with 4.00 mol oxygen gas.

62. Determine the number of moles of calcium oxide produced after the reaction is complete.

    ANSWER: 6.00 mol CaO

63. Determine the number of moles of calcium left over after the reaction is complete.

    ANSWER: 0 mol Ca (it is limiting)

64. Determine the number of moles of oxygen left over after the reaction is complete.

ANSWER: 1.00 mol O_2

65. You react 25.0 g hydrogen gas with 25.0 g oxygen gas. Determine the mass of water that can be produced from these reactants.

ANSWER: 28.2 g H_2O

66. Which of the following statements is always true concerning a reaction represented by the following balanced chemical equation?
$2C_2H_6(g) + 7O_2(g) \rightarrow 4CO_2(g) + 6H_2O(g)$
   a. If we have equal masses of C_2H_6 and O_2, there is no limiting reactant.
   b. If we have an equal number of moles of C_2H_6 and O_2, there is no limiting reactant.
   c. If we have more mass of C_2H_6, then O_2 must be the limiting reactant.
   d. If we have more mass of O_2, then C_2H_6 must be the limiting reactant.
   e. None of these statements (a-d) are true.

ANSWER: c.   If we have more mass of C_2H_6, then O_2 must be the limiting reactant.

67. Equal masses of hydrogen gas and oxygen gas are reacted to form water. Which substance is limiting?

ANSWER: Oxygen gas (O_2) is limiting.

68. Reacting 3.00 mol nitrogen gas with 6.00 mol of hydrogen gas will produce how many moles of ammonia according to the following balanced chemical equation?
$N_2(g) + 3H_2(g) \rightarrow 2NH_3(g)$
   a. 2.00 mol NH_3
   b. 3.00 mol NH_3
   c. 4.00 mol NH_3
   d. 6.00 mol NH_3
   e. 9.00 mol NH_3

ANSWER: c.   4.00 mol NH_3

69. Consider a reaction in which two reactants make one product (for example, consider the unbalanced A + B → C). You know the following:
      2.0 mol A (with an excess of B) can produce a maximum of 2.0 mol C
      3.0 mol B (with an excess of A) can produce a maximum of 4.0 mol C
If you react 2.0 mol A with 3.0 mol B, what is the maximum amount of C that can be produced?
   a. 2.0 mol
   b. 4.0 mol
   c. 5.0 mol
   d. 6.0 mol
   e. More information is needed to answer this question.

ANSWER: a.   2.0 mol

70. Consider the equation: $A + 5B \rightarrow 3C + 4D$. When equal masses of A and B are reacted, which is limiting?
   a. If the molar mass of A is greater than the molar mass of B, then A must be limiting.
   b. If the molar mass of A is less than the molar mass of B, then A must be limiting.
   c. If the molar mass of A is greater than the molar mass of B, then B must be limiting.
   d. If the molar mass of A is less than the molar mass of B, then B must be limiting.
   e. Neither reactant is limiting.

   ANSWER: d.   If the molar mass of A is less than the molar mass of B, then B must be limiting.

71. Which of the following reaction mixtures would produce the greatest amount of product (assuming each goes to completion)? Each involves the reaction represented by the balanced equation
   $2H_2(g) + O_2(g) \rightarrow 2H_2O(g)$
   a. 2 mol H_2 and 2 mol O_2
   b. 2 mol H_2 and 3 mol O_2
   c. 2 mol H_2 and 1 mol O_2
   d. 3 mol H_2 and 1 mol O_2
   e. Each would produce the same amount of product.

   ANSWER: e.   Each would produce the same amount of product.

72. Ammonia reacts with oxygen to form nitrogen dioxide and water according to the following equation.
   $4NH_3(g) + 7O_2(g) \rightarrow 4NO_2(g) + 6H_2O(g)$
   You react ammonia and oxygen, and at the end of the experiment you find that you produced 27.0 g of water, and have 8.52 g of ammonia left over. Determine the mass of oxygen reacted.

   ANSWER: 56.0 g O_2

73. Consider the following reaction:
   $2A + B \rightarrow 3C + D$
   3.0 mol A and 2.0 mol B react to form 4.0 mol C. What is the percent yield of this reaction?
   a. 50%
   b. 67%
   c. 75%
   d. 89%
   e. 100%

   ANSWER: d.   89%

74. In the reaction between CO and Fe_3O_4, the theoretical yield in an experiment is calculated to be 47.2 g Fe. When a careless chemistry student carries out the experiment, the actual yield is 42.9 g Fe. Calculate the percentage yield.

   ANSWER: 90.9%

75.   When NH_3 is prepared from 28 g N_2 and excess H_2, the theoretical yield of NH_3 is 34 g. When this reaction is carried out in a given experiment, only 30. g is produced. What is the percentage yield?

a.   6%
b.   12%
c.   14%
d.   82%
e.   88%

ANSWER: e.   88%

CHAPTER 10
Modern Atomic Theory

1. The distance between two successive peaks or troughs in a wave is called
   a. the frequency
   b. the wavelength
   c. the speed
   d. the amplitude
   e. none of these

   ANSWER: b.   the wavelength

2. The _____ indicates the number of waves that pass a given point per second.
   a. frequency
   b. speed
   c. wavelength
   d. amplitude
   e. none of these

   ANSWER: a.   frequency

3. The _____ indicates how fast a given peak of a wave moves through space.
   a. frequency
   b. speed
   c. wavelength
   d. amplitude
   e. none of these

   ANSWER: b.   speed

4. A packet of energy of electromagnetic radiation is called
   a. a wavelength
   b. a wave
   c. a photon
   d. a proton
   e. none of these

   ANSWER: c.   a photon

5. We usually use the term _____ for all forms of electromagnetic radiation.
   a. energy
   b. photons
   c. radiation
   d. light
   e. none of these

   ANSWER: d. light

6. The form of EMR that has more energy per photon than ultraviolet rays but less energy per photon than gamma rays is
   a. microwaves
   b. radio waves
   c. X-rays
   d. infrared rays
   e. none of these

   ANSWER: c. X-rays

7. The form of EMR that has less energy per photon than infrared rays but more energy per photon than radio waves is
   a. microwaves
   b. untraviolet
   c. gamma rays
   d. X-rays
   e. none of these

   ANSWER: a. microwaves

8. Which color of visible light has the most energy per photon?
   a. violet
   b. blue
   c. green
   d. yellow
   e. red

   ANSWER: a. violet

9. Which color of visible light has the least amount of energy per photon?
   a. violet
   b. blue
   c. green
   d. yellow
   e. red

   ANSWER: e. red

10. When an electron in the ground state absorbs energy, it goes to a(n) _____ state.
    a. excited
    b. lower
    c. frenetic
    d. ionic
    e. stable

    ANSWER: a.   excited

11. The energy levels of the hydrogen atom (and all atoms) are _____, meaning that only certain discrete energy levels are allowed.
    a. varied
    b. quantized
    c. ramp-like
    d. continuous
    e. two of these

    ANSWER: b.   quantized

12. The color of a polar bear's fur is
    a. white
    b. brown
    c. black
    d. yellow
    e. colorless

    ANSWER: e.   colorless

13. The _____ is a phenomenon that may be caused by the burning of fossil fuels, which increases the CO_2 content in the earth's atmosphere.
    a. acid rain
    b. greenhouse effect
    c. infrared radiation
    d. ozone problem
    e. none of these

    ANSWER: b.   greenhouse effect

14. _____ is a form of oxygen that protects us from high-energy radiation emitted by the sun.
    a. Ozone
    b. Carbon dioxide
    c. Ultraviolet light
    d. Methane
    e. None of these

    ANSWER: a.   Ozone

15. Which of the following is a reasonable criticism of the Bohr model of the atom?
    a. It makes no attempt to explain why the negative electron does not eventually fall into the positive nucleus.
    b. It does not adequately predict the line spectrum of hydrogen.
    c. It does not adequately predict the ionization energy of the valence electron(s) for elements other than hydrogen.
    d. It does not adequately predict the ionization energy of the first-energy-level electrons for one-electron species for elements other than hydrogen.
    e. It shows the electrons to exist outside the nucleus.

    ANSWER: c.   It does not adequately predict the ionization energy of the valence electron(s) for elements other than hydrogen.

16. The probability map for an electron is called
    a. an orbit
    b. a photon
    c. an orbital
    d. an electron configuration
    e. none of these

    ANSWER: c.   an orbital

17. As the principal energy level increases in an atom's orbitals, the average distance of an electron energy level from the nucleus _____.
    a. increases
    b. decreases
    c. stays the same
    d. varies
    e. none of these

    ANSWER: a.   increases

18. The shape of an *p* orbital is
    a. spherical
    b. dumbbell shaped
    c. donut shaped
    d. conical shaped
    e. none of these

    ANSWER: b.   dumbbell shaped

19. A given set of *f* orbitals consists of _____ orbital(s).
    a. 1
    b. 3
    c. 5
    d. 7
    e. 9

    ANSWER: d. 7

20. The maximum number of electrons allowed in each of the p orbitals is
    a. 2
    b. 4
    c. 8
    d. 18
    e. none of these

    ANSWER: a. 2

21. A given set of d orbitals consists of _____ orbital(s).
    a. 1
    b. 3
    c. 5
    d. 6
    e. none of these

    ANSWER: c. 5

22. The maximum number of electrons allowed in each of the d orbitals is
    a. 2
    b. 4
    c. 8
    d. 18
    e. 32

    ANSWER: a. 2

23. The maximum electron capacity of an f sublevel is
    a. 18
    b. 14
    c. 10
    d. 6
    e. 2

    ANSWER: b. 14

24. A d sublevel can hold a maximum of
    a. 5 electrons
    b. 10 electrons
    c. 14 electrons
    d. 32 electrons
    e. none of these

    ANSWER: b. 10 electrons

25. The maximum number of electrons allowed in the p sublevel of the third principal level is
    a. 1
    b. 2
    c. 3
    d. 6
    e. 8

    ANSWER: d. 6

26. The lowest energy level to allow *f* orbitals is the fourth energy level. T/F _____

    ANSWER: True

27. The maximum number of electrons allowed in the fourth energy level is
    a. 2
    b. 4
    c. 8
    d. 18
    e. 32

    ANSWER: e.   32

28. State the maximum number of electrons allowed in each.
    a. fourth principal energy level _____
    b. any *d* sublevel _____
    c. a 2*p* orbital _____

    ANSWER:
    a. 32
    b. 10
    c. 2

29. Which of the following is an incorrect designation for an atomic orbital?
    a. 1*s*
    b. 4*f*
    c. 3*s*
    d. 2*d*
    e. 2*p*

    ANSWER: d.   2*d*

30. What is the maximum number of electrons that can be put into each of the following subshells?
    a. 3*p* _____
    b. 6*s* _____
    c. 5*d* _____
    d. 7*f* _____

    ANSWER:
    a. 6
    b. 2
    c. 10
    d. 14

31. The number of *d* orbitals in the second principal energy level is
    a. 2
    b. 6
    c. 10
    d. 14
    e. none of these

    ANSWER: e.   none of these

32. Phosphorus has how many electrons in its outermost principal energy level?
    a. 1
    b. 2
    c. 3
    d. 5
    e. 15

    ANSWER: d. 5

33. The number of unpaired electrons in an nitrogen atom is
    a. 1
    b. 2
    c. 3
    d. 4
    e. 5

    ANSWER: c. 3

34. The electron configuration for the sulfur atom is
    a. $1s^2 2s^2 2p^6 3s^2 3p^2$
    b. $1s^2 2s^2 2p^6 3s^2 3p^4$
    c. $1s^2 2s^2 2p^6 3s^5$
    d. $1s^2 2s^2 2p^6 3s^2 3p^5$
    e. none of these

    ANSWER: b. $1s^2 2s^2 2p^6 3s^2 3p^4$

35. The electron configuration for the oxygen atom is
    a. $1s^2 2p^6$
    b. $[He]\, 2s^6$
    c. $[Ne]\, 2s^2 2p^4$
    d. $1s^2 2s^2 2p^4$
    e. none of these

    ANSWER: d. $1s^2 2s^2 2p^4$

36. $1s^2 2s^2 2p^6 3s^2 3p^5$ is the correct electron configuration for which of the following atoms?
    a. F
    b. S
    c. P
    d. Cl
    e. none of these

    ANSWER: d. Cl

37. The halogens (Group 7) contain how many valence electrons?
    a. 1
    b. 7
    c. 0
    d. 8
    e. none of these

    ANSWER: b. 7

38. The elements chlorine and iodine have similar chemical properties because they
    a. are both metals
    b. are in the same chemical period
    c. have the same number of electrons in their outer energy levels
    d. have the same number of stable isotopes
    e. none of these

    ANSWER: c.   have the same number of electrons in their outer energy levels

39. The alkaline earth metals have how many valence electrons?
    a. 8
    b. 7
    c. 3
    d. 2
    e. 1

    ANSWER: d. 2

40. Which of the following is the highest energy orbital for a silicon atom?
    a. 1s
    b. 2s
    c. 3s
    d. 3p
    e. 3d

    ANSWER: e. 3d

41. How many electrons are in the third principal energy level ($n = 3$) of one atom of Fe?
    a. 2
    b. 8
    c. 14
    d. 18
    e. none of these

    ANSWER: c. 14

42. The noble gases contain how many valence electrons?
    a. 1
    b. 7
    c. 0
    d. 8
    e. none of these

    ANSWER: d. 8

43. The number of electrons in the third sublevel of an iron atom is
    a. 3
    b. 6
    c. 8
    d. 26
    e. 56

    ANSWER: b.   6

44. All these atoms have seven electrons in their outermost energy levels *except*
    a. H
    b. F
    c. Cl
    d. Br
    e. I

    ANSWER: a.   H

45. The maximum number of electrons in the second principal energy level of an atom is
    a. 2
    b. 6
    c. 8
    d. 18
    e. 32

    ANSWER: c.   8

46. Which element has the fewest number of electrons in its valence shell?
    a. Cs
    b. Mg
    c. P
    d. O
    e. Br

    ANSWER: a.   Cs

47. How many *d* electrons are there in a iron atom?
    a. 2
    b. 3
    c. 6
    d. 26
    e. 56

    ANSWER: c.   6

48. Which one of the following atoms has a partly filled *d* sublevel?
    a. Ca
    b. Ni
    c. Zn
    d. As
    e. Ar

    ANSWER: b.   Ni

49. When moving down a group (family) in the periodic table, the number of valence electrons
    a. remains constant
    b. increases by 2 then 8 then 18 then 32
    c. doubles with each move
    d. decreases regularly
    e. changes in an unpredictable manner

    ANSWER: a.   remains constant

50. The Group 3 elements through the Group 8 elements form an area of the periodic table where the electron sublevels being filled are
    a. *p* orbitals
    b. *s* and *p* orbitals
    c. *d* orbitals
    d. *p* and *d* orbitals
    e. *f* orbitals

    ANSWER: a.   *p* orbitals

51. The yet undiscovered element with atomic number 113 would be a member of
    a. the halogens
    b. the transition elements
    c. the noble gases
    d. the Group 3 elements
    e. none of these

    ANSWER: d.   the Group 3 elements

52. $1s^2 2s^2 2p^6 3s^2 3p^6 4s^2 3d^6$ is the electron configuration for which of the following atoms?
    a. Ca
    b. Fe
    c. Cr
    d. Ar
    e. none of these

    ANSWER: b.   Fe

53. $1s^2 2s^2 2p^6 3s^2 3p^6 4s^2 3d^7$ is the electron configuration for which of the following atoms?
    a. Ca
    b. Fe
    c. Cr
    d. Ar
    e. Co

    ANSWER: e.   Co

54. The correct electron configuration for Mn is
    a. $1s^2 2s^2 2p^6 3s^2 3p^6 3d^7$
    b. $1s^2 2s^2 2p^6 3s^2 3p^6 4s^2 3d^6$
    c. $1s^2 2s^2 2p^6 2d^{10} 3s^2 3p^3$
    d. $1s^2 2s^2 2p^6 3s^2 3p^6 4s^2 3d^5$
    e. none of these

    ANSWER: d. $1s^2 2s^2 2p^6 3s^2 3p^6 4s^2 3d^5$

55. The electron configuration for manganese is
    a. $[Ar] 3d^7$
    b. $1s^2 2s^2 2p^6 3s^1 3d^6$
    c. $[Ar] 4s^2 3d^5$
    d. $1s^2 2s^2 2p^6 3s^2 3d^4$
    e. $[Ar] 4s^2 4p^5$

    ANSWER: c. $[Ar] 4s^2 3d^5$

56. Which of the following has the electron configuration $1s^2 2s^2 2p^6 3s^2 3p^6 4s^2 3d^5$?
    a. Cr
    b. Ca
    c. Mn
    d. Br
    e. none of these

    ANSWER: c. Mn

57. Which of the following atoms has the electron configuration $1s^2 2s^2 2p^6 3s^2 3p^6 4s^2 3d^1$?
    a. Sc
    b. Ca
    c. Sr
    d. Ar
    e. none of these

    ANSWER: a. Sc

58. The number of unpaired electrons in a cobalt atom is
    a. 2
    b. 3
    c. 5
    d. 7
    e. none of these

    ANSWER: b. 3

59. Which electron configuration indicates a transitional element?
    a. $1s^2 2s^2 2p^6 3s^1 3p^6$
    b. $1s^2 2s^2 2p^6 3s^2 3p^6 4s^2 3d^3$
    c. $1s^2 2s^2 2p^5$
    d. $1s^2 2s^2 2p^6 3s^2 3p^6 4s^2 3d^{10} 4p^2$
    e. none of these

    ANSWER: b. $1s^2 2s^2 2p^6 3s^2 3p^6 4s^2 3d^3$

60. The element with the electron configuration [Kr] $5s^2 4d^{10} 5p^3$ is
    a. As
    b. Sb
    c. Nb
    d. Pr
    e. none of these

    ANSWER: b.   Sb

61. How many of the following electron configurations for the species in their ground state are correct?
    I. Ca: $1s^2 2s^2 2p^6 3s^2 3p^6 4s^2$
    II. Mg: $1s^2 2s^2 2p^6 3s^1$
    III. V: [Ar] $3s^2 3d^3$
    IV. As: [Ar] $4s^2 3d^{10} 4p^3$
    V. P: $1s^2 2s^2 2p^6 3p^5$
    a. 1
    b. 2
    c. 3
    d. 4
    e. 5

    ANSWER: b.   2

62. The p block of elements contains the transition metals. T/F _____

    ANSWER: False

63. The 47th electron of silver, Ag, will be in a d orbital. T/F _____

    ANSWER: True

64. What element has the electron configuration $1s^2 2s^2 2p^6 3s^2 3p^6 4s^1 3d^{10}$?
    a. Zn
    b. Ga
    c. Cu
    d. Cd
    e. Ag

    ANSWER: c.   Cu

65. What element has the electron configuration $1s^22s^22p^63s^23p^64s^23d^8$?
    a. Ni
    b. Cu
    c. Pd
    d. Zn
    e. Ar

    ANSWER: a. Ni

66. What element has the electron configuration $1s^22s^22p^63s^23p^64s^1$?
    a. Rb
    b. Ca
    c. Sc
    d. K
    e. none of these

    ANSWER: d. K

67. What element has the electron configuration $1s^22s^22p^63s^23p^64s^23d^{10}4p^5$?
    a. Cl
    b. Se
    c. I
    d. Kr
    e. Br

    ANSWER: e. Br

68. What element has the electron configuration $1s^22s^22p^63s^23p^3$?
    a. N
    b. P
    c. S
    d. Al
    e. Cl

    ANSWER: b. P

69. What element has the electron configuration $1s^22s^22p^63s^23p^6$?
    a. Ar
    b. Cl
    c. Kr
    d. S
    e. none of these

    ANSWER: a. Ar

70. What element has the electron configuration
    $1s^22s^22p^63s^23p^64s^23d^{10}4p^65s^24d^{10}5p^66s^24f^{14}5d^{10}6p^2$?
    a. Ba
    b. Sn
    c. Pb
    d. Po
    e. none of these

    ANSWER: c. Pb

71. What element has the electron configuration $1s^22s^22p^63s^23p^64s^23d^{10}4p^65s^24d^{10}5p^66s^2$?
     a. Ba
     b. Sn
     c. Pb
     d. Po
     e. none of these

     ANSWER: a.   Ba

72. Write the electron configuration for Se.

     ANSWER: $1s^22s^22p^63s^23p^64s^23d^{10}4p^4$ or $[Ar]\,4s^23d^{10}4p^4$

73. Write the electron configuration for Co.

     ANSWER: $[Ar]\,4s^23d^7$

74. Write the electron configuration for Ca.

     ANSWER: $1s^22s^22p^63s^23p^64s^2$ or $[Ar]\,4s^2$

75. Write the electron configuration for Cd.

     ANSWER: $[Kr]\,5s^24d^{10}$

76. Write the electron configuration for Te.

     ANSWER: $[Kr]\,5s^24d^{10}5p^4$

77. Write the electron configuration for Cl.

     ANSWER: $[Ne]\,3s^23p^5$

78. Write the electron configuration for Al.

     ANSWER: $[Ne]\,3s^23p^1$

79. Write the electron configuration for B.

     ANSWER: $1s^22s^22p^1$

80. Write the electron configuration for Br.

     ANSWER: $[Ar]\,4s^23d^{10}4p^5$

81. Write the electron configuration for Rb.

     ANSWER: $[Kr]\,5s^1$

82. Write the electron configuration for Ba.

    ANSWER: [Xe] $6s^2$

83. Which of the following atoms has the largest atomic radius?
    a.   Na
    b.   Mg
    c.   Si
    d.   P
    e.   C

    ANSWER: a.   Na

84. Which of the following atoms has the highest ionization energy?
    a.   Na
    b.   Mg
    c.   Si
    d.   P
    e.   Cl

    ANSWER: e.   Cl

85. Which of the following atoms has the smallest atomic radius?
    a.   As
    b.   Sb
    c.   Bi
    d.   P
    e.   N

    ANSWER: e.   N

86. Which of the following has the largest atomic radius?
    a.   Na
    b.   Mg
    c.   P
    d.   N
    e.   O

    ANSWER: a.   Na

87. Which of the following has the highest ionization energy?
    a.   K
    b.   Ca
    c.   C
    d.   N
    e.   O

    ANSWER: e.   O

88. Which of the following has the smallest atomic radius?
    a. N
    b. F
    c. Br
    d. Cl
    e. S

    ANSWER: b. F

89. Which of the following atoms has the highest ionization energy?
    a. Al
    b. Si
    c. P
    d. As
    e. Sb

    ANSWER: c. P

90. Metal atoms tend to _____ electrons and form _____ ions.
    a. lose; positive
    b. lose; negative
    c. gain; negative
    d. gain; positive
    e. share; neutral

    ANSWER: a. lose; positive

91. Which of the following exhibits the correct orders for both atomic radius and ionization energy, respectively?
    a. S, O, F, and F, O, S
    b. F, S, O, and O, S, F
    c. S, F, O, and S, F, O
    d. F, O, S, and S, O, F
    e. none of these

    ANSWER: d. F, O, S, and S, O, F

92. Of the metals in Group 1, which has the highest ionization energy?
    a. Cs
    b. Rb
    c. K
    d. Na
    e. Li

    ANSWER: e. Li

93. Order the elements S, Cl, and F in terms of increasing ionization energy.
    a. S, Cl, F
    b. Cl, F, S
    c. F, S, Cl
    d. F, Cl, S
    e. S, F, Cl

    ANSWER: a. S, Cl, F

94. Order the elements S, Cl, and F in terms of increasing atomic radii.
    a. S, Cl, F
    b. Cl, F, S
    c. F, S, Cl
    d. F, Cl, S
    e. S, F, Cl

    ANSWER: d. F, Cl, S

95. Which has the larger atomic radius, S or Si?

    ANSWER: Si

96. Which has the higher ionization energy, Rb or Cs?

    ANSWER: Rb

97. Which has the higher ionization energy, K or Br?

    ANSWER: Br

98. Which of the following properties generally increases as we go from the lower left to the upper right of the periodic table?
    a. ionization energy
    b. atomic size
    c. atomic number
    d. number of electron shells
    e. none of these

    ANSWER: a. ionization energy

99. What are the general trends in ionization energy
    a. across a period?
    b. down a group (family)?

    ANSWER:
    a. increases
    b. decreases

100. How does magnesium compare with sodium in terms of the following properties?
    a. atomic size
    b. number of outer shell electrons
    c. ionization energy
    d. formula of the bromide salt

    ANSWER:
    a. smaller
    b. larger
    c. larger
    d. $MgBr_2$; NaBr

CHAPTER 11
Chemical Bonding

1. When electrons are shared unequally, chemists characterize these types of bonds as
_____.
   a. polar covalent
   b. ionic
   c. pure covalent
   d. unbalanced
   e. none of these

   ANSWER: a.   polar covalent

2. Chemical bonds formed by the attraction of oppositely charged ions are called
   a. covalent bonds
   b. magnetic bonds
   c. coordinate bonds
   d. ionic bonds
   e. none of these

   ANSWER: d.   ionic bonds

3. Atoms with greatly differing electronegativity values are expected to form
   a. no bonds
   b. polar covalent bonds
   c. nonpolar covalent bonds
   d. triple bonds
   e. ionic bonds

   ANSWER: e.   ionic bonds

4. Metals typically have _____ electronegativity values.
   a. high
   b. low
   c. negative
   d. no
   e. two of these

   ANSWER: b.   low

5. Nonmetal elements typically have _____ electronegativities.
    a. low
    b. high
    c. neutral
    d. strong
    e. none of these

    ANSWER: b. high

6. In general, a larger atom has a smaller electronegativity. T/F _____

    ANSWER: False

7. The electron pair in a C--F bond could be considered
    a. closer to C because carbon has a larger radius and thus exerts greater control over the shared electron pair
    b. closer to F because fluorine has a higher electronegativity than carbon
    c. closer to C because carbon has a lower electronegativity than fluorine
    d. an inadequate model because the bond is ionic
    e. centrally located directly between the C and F

    ANSWER: b. closer to F because fluorine has a higher electronegativity than carbon

8. What is the correct order of the following bonds in terms of decreasing polarity?
    a. N--Cl, P--Cl, As--Cl
    b. P--Cl, N--Cl, As--Cl
    c. As--Cl, N--Cl, P--Cl
    d. P--Cl, As--Cl, N--Cl
    e. As--Cl, P--Cl, N--Cl

    ANSWER: e. As--Cl, P--Cl, N--Cl

9. Which of the following bonds would be the most polar without being considered ionic?
    a. Mg--O
    b. C--O
    c. O--O
    d. Si--O
    e. N--O

    ANSWER: d. Si--O

10. Which of the following bonds would be the least polar yet still be considered polar covalent?
    a. Mg--O
    b. C--O
    c. O--O
    d. Si--O
    e. N--O

    ANSWER: e. N--O

11. An N--F bond is expected to be more polar than an O--F bond. T/F _____

    ANSWER: True

12. Carbon dioxide has _____ bonds.
    a. ionic
    b. covalent
    c. polar covalent
    d. magnetic
    e. none of these

    ANSWER: c.   polar covalent

Use the following choices to classify each of the following molecules.
    a. ionic
    b. covalent

13. CO_2 _____

    ANSWER: b.   covalent

14. OCl_2 _____

    ANSWER: b.   covalent

15. K_2O _____

    ANSWER: a.   ionic

16. $MgCl_2$ _____

    ANSWER: a.   ionic

17. NaBr _____

    ANSWER: a.   ionic

18. NH_3 _____

    ANSWER: b.   covalent

19. $CoCl_2$ _____

    ANSWER: a.   ionic

20. CH_4 _____

    ANSWER: b.   covalent

21. SF_4 _____

    ANSWER: b.   covalent

22. $AlCl_3$ _____

    ANSWER: a.   ionic

Use the following choices to classify the bonds in each of the following molecules.
  a. nonpolar
  b. polar

23. S_8 _____.

  ANSWER: a. nonpolar

24. CF_4 _____

  ANSWER: a. nonpolar

25. H_2S _____

  ANSWER: b. polar

26. I_2 _____

  ANSWER: a. nonpolar

27. CO _____

  ANSWER: b. polar

28. Cl_2 _____

  ANSWER: a. nonpolar

29. Which of the following compounds contains an ionic bond?
  a. $HCl(g)$
  b. NaCl
  c. CCl_4
  d. SO_2
  e. O_2

  ANSWER: b. NaCl

30. Which of the following compounds contains one or more covalent bonds?
  a. NaCl
  b. CaO
  c. CO_2
  d. Cs_2O
  e. $BaBr_2$

  ANSWER: c. CO_2

31. The most electronegative element is
     a.   He
     b.   F
     c.   At
     d.   Cs
     e.   O

     ANSWER: b.   F

32. In the hydrogen chloride molecule, the atoms are held together by a(n)
     a.   ionic bond
     b.   polar covalent bond
     c.   nonpolar bond
     d.   double bond
     e.   none of these

     ANSWER: b.   polar covalent bond

33. Which of these is *not* an ionic compound?
     a. K_2CO_3
     b.   HCl
     c.   NaSCN
     d. NH_4I
     e. $MgCl_2$

     ANSWER: b.   HCl

34. Which of the following elements has the lowest electronegativity?
     a.   Na
     b.   Rb
     c.   Ca
     d.   S
     e.   Cl

     ANSWER: b.   Rb

35. Which of these elements has the highest electronegativity?
     a.   Y
     b.   I
     c.   Sb
     d.   Sr
     e.   In

     ANSWER: b.   I

36. Which of the following elements has the lowest electronegativity?
     a.   H
     b.   S
     c.   Cl
     d.   Ca
     e.   Ba

     ANSWER: e.   Ba

37. Which of the following has nonpolar bonds?
    a. H_2S
    b. HCl
    c. Br_2
    d. OF_2
    e. All are nonpolar.

    ANSWER: c. Br_2

38. Which of the following contains only nonpolar bonds?
    a. CH_4
    b. HCl
    c. H_2O
    d. Mg_3N_2
    e. Cl_2

    ANSWER: e. Cl_2

39. Which of the following has only nonpolar covalent bonds?
    a. N_2
    b. CO
    c. HI
    d. CCl_4
    e. NaCl

    ANSWER: a. N_2

40. Which of the following is the most electronegative?
    a. Na
    b. Mg
    c. Si
    d. P
    e. Cl

    ANSWER: e.   Cl

41. Which of the following atoms has the greatest electronegativity?
    a. Na
    b. Rb
    c. Cl
    d. Se

    ANSWER: c.   Cl

42. Which of the following has primarily ionic bonding?
    a. N_2O_3
    b. Na_2O
    c. CO_2
    d. CCl_4
    e. none of these

    ANSWER: b. Na_2O

43. The number of polar covalent bonds in NH_3 is
    a. 1
    b. 2
    c. 3
    d. 4
    e. none of these

    ANSWER: c. 3

44. If atom X forms a diatomic molecule with itself, the bond is
    a. ionic
    b. polar covalent
    c. nonpolar covalent
    d. polar coordinate covalent
    e. none of these

    ANSWER: c. nonpolar covalent

45. Which of the following has ionic bonding?
    a. K_2S
    b. SO_3
    c. CS_2
    d. SiH_4
    e. none of these

    ANSWER: a. K_2S

46. Arrange the following elements in order of increasing electronegativity (from the smallest to the largest).
    a. N < C < Be < F
    b. C < F < Be < N
    c. F < N < C < Be
    d. Be < C < N < F
    e. C < N < F < Be

    ANSWER: d. Be < C < N < F

47. The water molecule has a dipole moment. T/F _____

    ANSWER: True

48. When a molecule has a center of positive charge and a center of negative charge, it is said to have a _____.
    a. magnetic attraction
    b. diatomic bond
    c. double bond
    d. polyatomic ion
    e. dipole moment

    ANSWER: e. dipole moment

49. One of the most important characteristics of the water molecule is its _____, which allows it to surround and attract both positive and negative ions.
    a. polarity
    b. strength
    c. magnetism
    d. fluidity
    e. stability

    ANSWER: a. polarity

50. Which element or ion listed below has the electron configuration $1s^2 2s^2 2p^6 3s^2 3p^6$?
    a. Cl
    b. Br$^-$
    c. Se
    d. Ca^{2+}
    e. two of these

    ANSWER: d. Ca^{2+}

51. Which element or ion listed below has the electron configuration $1s^2 2s^2 2p^6 3s^2 3p^6$?
    a. Cl
    b. Br$^-$
    c. Se
    d. Ca^{2+}
    e. two of these

    ANSWER: e. two of these

52. Which element or ion listed below has the electron configuration $1s^2 2s^2 2p^6$?
    a. Na$^+$
    b. Al^{3+}
    c. F$^-$
    d. Ne
    e. all of these

    ANSWER: e. all of these

53. The F$^-$ and O^{2-} ions have the same electron configuration. T/F _____

    ANSWER: True

54. A phosphorus atom needs to gain _____ electrons to achieve a noble gas configuration.
    a. 2
    b. 3
    c. 4
    d. 5
    e. 6

    ANSWER: b. 3

55. Which of the following species would be expected to have the lowest ionization energy?
    a. Br^-
    b. Kr
    c. Se^{2-}
    d. Sr^{2+}
    e. Rb^+

    ANSWER: c. Se^{2-}

56. Which of the following has the largest radius?
    a. S^{2-}
    b. Cl^-
    c. Ar
    d. K^+
    e. Ca^{2+}

    ANSWER: a. S^{2-}

57. Which of the following has the smallest radius?
    a. S^{2-}
    b. Cl^-
    c. Ar
    d. K^+
    e. Ca^{2+}

    ANSWER: e. Ca^{2+}

58. Which of the following ions has the same electron configuration as an argon atom?
    a. Br^-
    b. S^{3-}
    c. P^{3+}
    d. K^+
    e. Ca^+

    ANSWER: d. K^+

59. $1s^2 2s^2 2p^6 3s^2 3p^6$ is the electron configuration for which one of the following ions?
   a. S^{2-}
   b. Ca^+
   c. Na^+
   d. F^-
   e. none of these

   ANSWER: a. S^{2-}

60. $1s^2 2s^2 2p^6 3s^2 3p^6$ is the electron configuration for which one of the following ions?
   a. S^-
   b. Ca^{2+}
   c. Na^+
   d. F^-
   e. none of these

   ANSWER: b. Ca^{2+}

61. The electron configuration for the bromide ion is identical to that of
   a. Br
   b. Kr
   c. K
   d. I^-
   e. none of these

   ANSWER: b. Kr

62. The electron configuration for Ca^{2+} is identical with
   a. Ne
   b. Kr
   c. Ca
   d. Ar

   ANSWER: d. Ar

63. Which one of the following species has the same electron configuration as an atom of argon?
   a. S^-
   b. Cl^{2-}
   c. K
   d. Ca^{2+}
   e. Kr

   ANSWER: d. Ca^{2+}

64. Magnesium reacts with sulfur to form
    a. MgS
    b. MgS_2
    c. Mg_2S
    d. Mg_2S_3
    e. none of these

    ANSWER: a. MgS

65. Calcium reacts with fluorine to form
    a. CaF
    b. CaF_2
    c. Ca_2F
    d. Ca_2F_3
    e. none of these

    ANSWER: b. CaF_2

66. Which of the following compounds is the product of the reaction $Mg + O_2$?
    a. MgO
    b. MgO_2
    c. MgO_3
    d. Mg_2O
    e. Mg_2O_3

    ANSWER: a. MgO

67. Which of the following is the product of the reaction $Al + O_2$?
    a. AlO
    b. AlO_2
    c. AlO_3
    d. Al_3O_2
    e. Al_2O_3

    ANSWER: e. Al_2O_3

68. When they react chemically, the alkali metals (Group 1)
    a. gain one electron
    b. gain seven electrons
    c. gain or lose seven electrons
    d. lose one electron

    ANSWER: d. lose one electron

69. The formula of the compound formed in the reaction between lithium and sulfur is
    a. LiS
    b. LiS_2
    c. Li_2S_3
    d. Li_2S
    e. none of these

    ANSWER: d. Li_2S

70. Which element listed below has the electron configuration $1s^22s^22p^63s^23p^4$?
    a. Se
    b. O
    c. P
    d. S
    e. none of these

    ANSWER: d. S

71. Which element or ion listed below has the electron configuration $1s^22s^22p^63s^23p^6$?
    a. S
    b. Ne
    c. Cl
    d. S^{2-}
    e. none of these

    ANSWER: d. S^{2-}

72. Which element or ion listed below has the electron configuration $1s^22s^22p^63s^23p^6$?
    a. Ca^+
    b. Na^+
    c. K^+
    d. Cl
    e. none of these

    ANSWER: c. K^+

73. Write the electron configuration for Br^-.

    ANSWER: $1s^22s^22p^63s^23p^64s^23d^{10}4p^6$ or $[Ar]\,4s^23d^{10}4p^6$

74. Write the electron configuration for Sr^{2+}.

    ANSWER: $1s^22s^22p^63s^23p^64s^23d^{10}4p^6$

75. Write the electron configuration for Cl^-.

    ANSWER: $1s^22s^22p^63s^23p^6$

76. Write the electron configuration for Al^{3+}.

ANSWER: $1s^22s^22p^6$

77. Complete the table by giving the predicted formulas of the compounds formed between the elements listed.

	Br	S
Na	_____	_____
Mg	_____	_____
Al	_____	_____

ANSWER:

	Br	S
Na	NaBr	Na_2S
Mg	$MgBr_2$	MgS
Al	$AlBr_3$	Al_2S_3

78. Draw the Lewis electron structure for the silicon atom.

ANSWER:

$\cdot \overset{\displaystyle ..}{\underset{\displaystyle .}{Si}}$

79. Draw the Lewis electron structure for the sulfur atom.

ANSWER:

$\overset{\displaystyle ..}{\underset{\displaystyle .}{:S}} \cdot$

80. Draw the Lewis electron structure for the sulfide ion.

ANSWER:

$\left[: \overset{\displaystyle ..}{S} : \right]^{2-}$

81. Draw the Lewis electron structure for the chlorine atom.

ANSWER:

$: \overset{\displaystyle ..}{\underset{\displaystyle .}{Cl}} :$

82. How many lone pairs of electrons are in the Lewis structure for ammonia, NH_3?
    a. 0
    b. 1
    c. 2
    d. 3
    e. 4

ANSWER: b. 1

83. The Lewis structure for which of the following contains the greatest number of lone pairs of electrons?
    a. CH_4
    b. HF
    c. F_2
    d. H_2O
    e. H_2

    ANSWER: c. F_2

84. Choose the correct Lewis structure for the OH⁻ ion.

    a. $\begin{bmatrix} \ddot{O} & \cdot & H \end{bmatrix}^-$

    b. $\begin{bmatrix} O & : & H \end{bmatrix}^-$

    c. $\begin{bmatrix} :\ddot{O}: & H \end{bmatrix}^-$

    d. $\begin{bmatrix} \cdot\ddot{O}: & H \end{bmatrix}^-$

    ANSWER:

    c. $\begin{bmatrix} :\ddot{O}: & H \end{bmatrix}^-$

85. Choose the correct Lewis structure for the NH₄· ion.

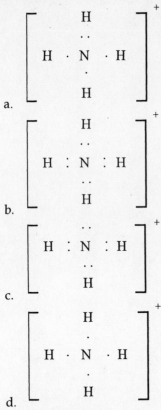

a.

b.

c.

d.

e. none of these

ANSWER:

$$\left[\begin{array}{c} H \\ \cdot\cdot \\ H : N : H \\ \cdot\cdot \\ H \end{array}\right]^{+}$$

b.·

86. Draw the Lewis electron structure for the Cl_2 molecule.

ANSWER:

$$: \overset{\cdot\cdot}{\underset{\cdot\cdot}{Cl}} - \overset{\cdot\cdot}{\underset{\cdot\cdot}{Cl}} :$$

87. Draw the Lewis electron structure for the HI molecule.

ANSWER:

$$H - \overset{\cdot\cdot}{\underset{\cdot\cdot}{I}} :$$

88. Draw the Lewis electron structure for the AsH_3 molecule.

ANSWER:

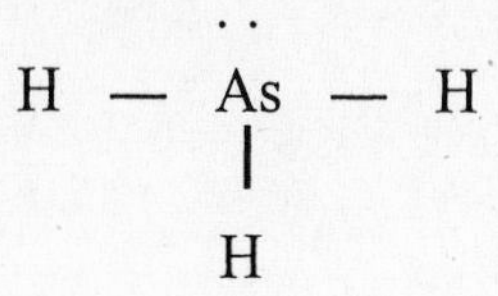

$$H - \overset{..}{As} - H$$
$$|$$
$$H$$

89. Draw the Lewis electron structure for the H_2Te molecule.

ANSWER:

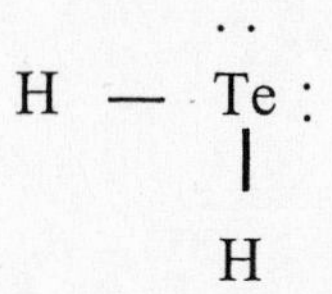

$$H - \overset{..}{Te} :$$
$$|$$
$$H$$

90. Draw the Lewis structure for NI_3.

ANSWER:

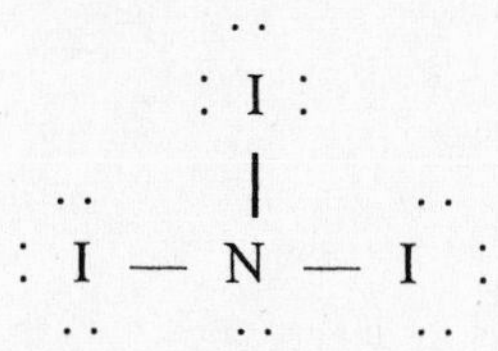

$$: \overset{..}{I} :$$
$$|$$
$$: \overset{..}{\underset{..}{I}} - \underset{..}{N} - \overset{..}{\underset{..}{I}} :$$

91. Draw the Lewis structure for SiH_4.

ANSWER:
$$H$$
$$|$$
$$H - Si - H$$
$$|$$
$$H$$

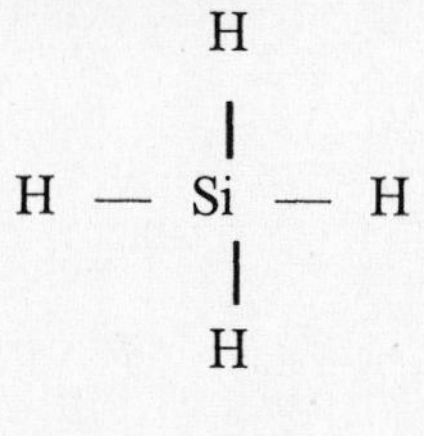

92. Draw the Lewis structure for CCl_4.

ANSWER:
$$: \overset{..}{Cl} :$$
$$|$$
$$: \overset{..}{\underset{..}{Cl}} - C - \overset{..}{\underset{..}{Cl}} :$$
$$|$$
$$: \overset{..}{\underset{..}{Cl}} :$$

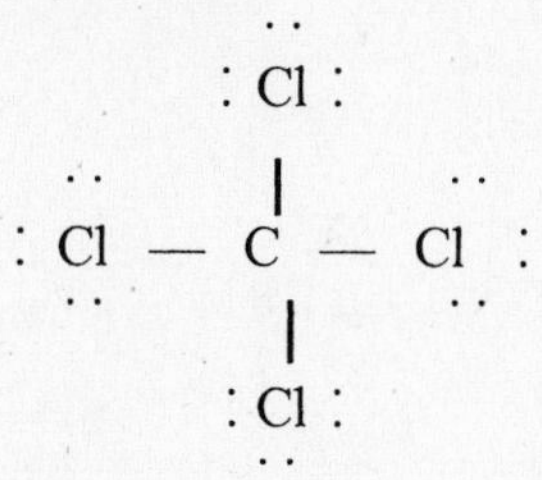

93. Draw the Lewis structure for Na_2O.

ANSWER:

$[Na]^+$
$[Na]^+$ $\left[:\overset{..}{\underset{..}{O}}: \right]^{2-}$

94. Draw the Lewis structure for KCN.

ANSWER:

$[K]^+$ $[:C \equiv N:]^-$

95. Draw the Lewis structure for CO.

ANSWER:

$:C \equiv O:$

96. Draw the Lewis structure for N_2.

ANSWER:

$:N \equiv N:$

97. How many resonance structures are there for SO_3?

ANSWER: 3

98. How many of the following will have Lewis structures with multiple bonds?
CO, CO_2, CO_3^{2-}, N_2, O_2
a. 1
b. 2
c. 3
d. 4
e. 5

ANSWER: e. 5

99. Draw the Lewis structure for HCN.

ANSWER:

$H — C \equiv N:$

100.   Draw the Lewis structure for $HCCl_3$.

ANSWER:

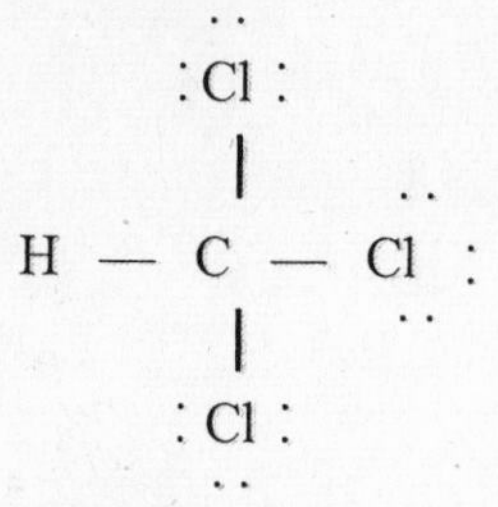

$$
\begin{array}{c}
\ddot{}\\
: \ddot{Cl} : \\
| \\
H - C - \ddot{Cl} : \\
| \\
: \ddot{Cl} :
\end{array}
$$

101.   Which of the following has a double bond?
a. H_2O
b. C_2H_2
c. C_2H_4
d. CN^-
e.   none of these

ANSWER: c. C_2H_4

102.   Which of the following has a triple bond?
a. CH_4
b. CO
c. SO_2
d. NO_3^-
e.   none of these

ANSWER: b. CO

103.   Which of the following has a double bond?
a. H_2O
b. NH_3
c. O_2
d. CO
e. H_2S

ANSWER: c. O_2

104.   Which of the following has a triple bond?
a. H_2O
b. NH_3
c. O_2
d. CO
e. H_2S

ANSWER: d. CO

Use the following choices to describe the molecular structure of each of the following molecules or ions.

    a. linear
    b. trigonal planar
    c. tetrahedral
    d. pyramidal
    e. V-shaped

105. CH_4

    ANSWER: c.   tetrahedral

106. SO_2

    ANSWER: e.   V-shaped

107. PF_3

    ANSWER: d.   pyramidal

108. OCl_2

    ANSWER: e.   V-shaped

Select the correct molecular structure for the given species from the choices below.

    a. linear
    b. bent
    c. pyramidal
    d. tetrahedral
    e. none of these

109. CO_2

    ANSWER: a.   linear

110. HCl

    ANSWER: a.   linear

111. NH_3

    ANSWER: c.   pyramidal

Consider the molecule SO_2. Answer the following.

112. What are the angles of the S-O bonds?

    ANSWER: 120°

113. What is the electron arrangement around the central atom?

    ANSWER: trigonal planar

114. What is the molecular geometry around the central atom?

    ANSWER: bent or V-shaped

115. How many lone pairs of electrons are around the central atom?

    ANSWER: one lone pair of electrons

116. Is the molecule polar or nonpolar?

    ANSWER: polar

CHAPTER 12

Gases

1. Convert 3.6×10^2 atm to torr.
   a. 13,000 torr
   b. 25 torr
   c. 0.47 torr
   d. 53,000 torr
   e. 270,000 torr

   ANSWER: e.   270,000 torr

2. Convert 973 torr to psi.
   a. 103 psi
   b. 18.8 psi
   c. 50400 psi
   d. 66.2 psi
   e. 1.28 psi

   ANSWER: b.   18.8 psi

3. The air in the inner tube of the tire of a racing bike has a pressure of 115 psi. Convert this pressure to atm.
   a. 0.151 atm
   b. 7.83 atm
   c. 1690 atm
   d. 32.6 atm
   e. 115 atm

   ANSWER: b.   7.83 atm

Perform the following conversions of pressure units:

4. 1.13 atm = _____ torr
   a. 859
   b. 430.
   c. 653
   d. 937
   e. 798

   ANSWER: a.   859

170

5. 168 torr = _____ atm
    a. 0.442
    b. 136
    c. 0.221
    d. 0.802
    e. 243

    ANSWER: c. 0.221

6. 5.0×10^9 Pa = _____ atm
    a. 9.8×10^4
    b. 2.5×10^4
    c. 1.7×10^5
    d. 4.3×10^4
    e. 4.9×10^4

    ANSWER: e. 4.9×10^4

7. 263 kPa = _____ Pa
    a. 0.263
    b. 5.26×10^5
    c. 2.63
    d. 2.63×10^5
    e. 2.63×10^4

    ANSWER: d. 2.63×10^5

8. 0.451 atm = _____ torr
    a. 248
    b. 553
    c. 0.683
    d. 343
    e. 395

    ANSWER: d. 343

9. 1.6×10^5 torr = _____ atm
    a. 3.2×10^2
    b. 2.1×10^2
    c. 2.1×10^3
    d. 3.2×10^3
    e. 4.6×10^2

    ANSWER: b. 2.1×10^2

10. Consider a gas at 1.00 atm in a 5.00 L container at 20°C. What pressure does the gas exert when transferred to a volume of 2.00 L at 43°C?

    ANSWER: 2.70 atm

11. You are holding two helium balloons, a large and a small balloon. How do the pressures of the helium compare?
    a. The pressure in the large balloon is greater than the pressure in the small balloon. This is best explained by the fact that there must be more moles of gas in the large balloon, thus greater pressure.
    b. The pressure in the small balloon is greater than the pressure in the large balloon. This is best explained by the fact that as the volume of a container decreases, the pressure of the gas increases.
    c. The pressure in the large balloon is greater than the pressure in the small balloon. The fact that there is a greater pressure in the balloon is why the volume is larger; as the particles push more on the inside, the volume increases.
    d. The pressures are essentially the same and less than atmospheric pressure.
    e. The pressures are essentially the same and equal to atmospheric pressure.

    ANSWER: e.   The pressures are essentially the same and equal to atmospheric pressure.

12. Consider a sample of gas in a container on a comfortable spring day in Chicago, IL. The Celsius temperature suddenly doubles, and you transfer the gas to a container with twice the volume of the first container. If the original pressure was 12 atm, what is a good estimate for the new pressure?
    a. 3 atm
    b. 5.5 atm
    c. 6.4 atm
    d. 12 atm
    e. 15 atm

    ANSWER: c.   6.4 atm

13. A certain balloon will pop if it expands to a volume greater than 16.0 L. The balloon is currently filled with air at a volume of 8.0 L. You heat the balloon such that the temperature measured in Celsius doubles. Which of the following best describes what happens?
    a. The balloon will expand to a volume greater than 16.0 L and pop.
    b. The balloon will expand to a volume less than 16.0 L and not pop.
    c. The balloon will expand to a volume of 16.0 L and pop.
    d. The balloon will expand to a volume of 16.0 L and not pop.
    e. The volume of the balloon will remain the same but the pressure will increase.

    ANSWER: b.   The balloon will expand to a volume less than 16.0 L and not pop.

14. A 15.0 g sample of a hydrocarbon is placed in a balloon at 1.00 atm and 25°C and the volume of the balloon is 12.2 L. The hydrocarbon is 79.89% carbon and 20.11% hydrogen by mass. Determine the molecular formula of the hydrocarbon.

    ANSWER: C_2H_6

15. Consider a steel container filled with 40.0 g of helium gas and 40.0 g of argon gas. What is the ratio of pressures that each gas exerts (helium:argon)?
    a. 1:1
    b. 10:1
    c. 1:10
    d. 9:1
    e. More information is needed to answer this question.

    ANSWER: b.   10:1

16. You are playing with a helium balloon on a typically warm California day. Suddenly, the Celsius temperature doubles. Which of the following is true?
    a. The volume of the balloon will double.
    b. The volume of the balloon will slightly increase.
    c. The pressure inside the balloon will double.
    d. The volume of the balloon will decrease.
    e. The actual temperatures are needed to answer this question.

    ANSWER: b. The volume of the balloon will slightly increase.

17. Determine the pressure exerted by 1.80 mol of gas in a 2.92 L container at 32°C.
    a. 1.62 atm
    b. 8.57 atm
    c. 15.4 atm
    d. 22.4 atm
    e. 495 atm

    ANSWER: c. 15.4 atm

18. You have a sample of argon gas at a certain pressure, volume, and temperature. You double the volume, double the number of moles of argon, and double the Kelvin temperature. How does the final pressure (P_f) compare to the original pressure (P_o)? $P_f =$
    a. $(1/8)P_o$
    b. $(1/2)P_o$
    c. $2P_o$
    d. $4P_o$
    e. $8P_2$

    ANSWER: c. $2P_o$

19. Calculate the density of neon gas at 1.00 atm and 25.0°C.
    a. 0.825 g/L
    b. 9.84 g/L
    c. 20.2 g/L
    d. 22.4 g/L
    e. None of these.

    ANSWER: a. 0.825 g/L

20. What is the volume of a helium balloon that contains 2.50 mol helium at 27°C and 1.10 atm?
    a. 5.04 L
    b. 22.4 L
    c. 34.8 L
    d. 56.0 L
    e. 61.5 L

    ANSWER: d. 56.0 L

21. A sample of a gas in a container fitted with a piston has a temperature above 0°C. The Celsius temperature is doubled. What is true about the ratio of final volume to initial volume for the gas?
    a. It is 1:1.
    b. It is 2:1.
    c. It is 1:2.
    d. It is greater than 2:1.
    e. It is less than 2:1.

    ANSWER: e.   It is less than 2:1.

22. If the temperature of an ideal gas is raised from 100°C to 200°C, while the pressure remains constant, the volume
    a. doubles
    b. remains the same
    c. goes to 1/2 the original volume
    d. increases by a factor of 100
    e. none of these

    ANSWER: e.   none of these

23. Which of the following will give a graph with a straight line and a *y*-intercept of 0?
    a. volume vs. 1/temperature (°C)
    b. volume vs. temperature (°C)
    c. volume vs. temperature (K)
    d. volume vs. 1/temperature (K)
    e. none of these

    ANSWER: c.   volume vs. temperature (K)

24. You transfer a sample of a gas at 17°C from a volume of 5.67 L and 1.10 atm to a container at 37°C that has a pressure of 1.10 atm. What is the new volume of the gas?
    a. 2.61 L
    b. 5.90 L
    c. 5.30 L
    d. 12.34 L
    e. none of these

    ANSWER: b.   5.90 L

25. A sample of helium gas occupies 12.4 L at 23°C and 0.956 atm. What volume will it occupy at 40°C and 0.956 atm?
    a. 7.13 L
    b. 11.7 L
    c. 21.6 L
    d. 13.1 L
    e. none of these

    ANSWER: d.   13.1 L

26. Gaseous chlorine is held in two separate containers at identical temperature and pressure. The volume of container 1 is 1.30 L, and it contains 6.70 mol of the gas. The volume of container 2 is 2.20 L. How many moles of the gas are in container 2?
    a. 11.3 mol
    b. 19.2 mol
    c. 0.427 mol
    d. 3.96 mol
    e. none of these

    ANSWER: a. 11.3 mol

27. Two moles of gas A spontaneously convert to three moles of gas B in a container where the temperature and pressure are held constant. The sample originally took up 10.2 L of volume. What is the new volume of the products?
    a. 0.189 L
    b. 6.73 L
    c. 12.3 L
    d. 1.15 L
    e. 15.3 L

    ANSWER: e. 15.3 L

28. If temperature and pressure are held constant, the volume and number of moles of a gas are
    a. independent of each other
    b. directly proportional
    c. inversely proportional
    d. equal
    e. not enough information given

    ANSWER: b. directly proportional

29. A sample of an ideal gas containing 0.954 mol is collected at 742 torr pressure and 31°C. Calculate the volume.

    ANSWER: 24.4 L

30. A gas originally occupying 10.1 L at 0.925 atm and 25°C is changed to 12.2 L at 625 torr. What is the new temperature?

    ANSWER: 47°C

31. One mole of CO_2 at STP will occupy
    a. 1.0 L
    b. 22.4 L
    c. 44 L
    d. 44 g
    e. 24.5 L

    ANSWER: b. 22.4 L

32. Which of the following statements is true of 19.0 g of $F_2(g)$ at STP?
    a. It contains 6.02×10^{23} molecules.
    b. It contains the same number of molecules as 1/2 mol of $O_2(g)$ at STP.
    c. It occupies a volume of 22.4 L.
    d. It only exists in the form of ions.
    e. none of the above

    ANSWER: b.   It contains the same number of molecules as 1/2 mol of $O_2(g)$ at STP.

33. You are holding two balloons, an orange balloon and a blue balloon. The orange balloon is filled with neon (Ne) gas, and the blue balloon is filled with argon (Ar) gas. The orange balloon has twice the volume of the blue balloon. Which of the following best represents the mass ratio of Ne:Ar in the balloons?
    a. 1:1
    b. 1:2
    c. 2:1
    d. 1:3
    e. 3:1

    ANSWER: a.   1:1

34. You are holding two balloons of the same volume. One balloon contains 1.0 g helium. The other balloon contains neon. What is the mass of neon in the balloon?
    a. 0.20 g
    b. 1.0 g
    c. 4.0 g
    d. 5.0 g
    e. 20. g

    ANSWER: d.   5.0 g

35. You fill a balloon with 10.0 g of N_2 gas. You wish to add 10.0 g of another gas to make the balloon more than twice as large as it is with only the N_2 (that is, more than twice the original volume). Which gas should you add (assume constant temperature)?
    a. O_2
    b. CO_2
    c. CO
    d. He
    e. You cannot make the balloon more than twice as large with 10.0 g of any gas.

    ANSWER: d.   He

36. A 12.0-g sample of helium gas occupies a volume of 25.0 L at a certain temperature and pressure. What volume does a 24.0-g sample of neon gas occupy at these conditions of temperature and pressure?
    a. 9.92 L
    b. 14.8 L
    c. 22.4 L
    d. 25.0 L
    e. 50.0 L

    ANSWER: a.   9.92 L

37. Suppose a balloon has a maximum volume of 7.00 L. Also suppose each time you blow into a balloon, you expel 0.045 mol of air. How many times can you blow into the balloon before the balloon pops. Assume atmospheric conditions of 1.0 atm and 22°C and that your breath is room temperature.

    ANSWER: We can blow into the balloon 6 times. It will pop during the 7th time.

38. You have a certain mass of helium gas in a rigid steel container. You add the same mass of neon gas to this container. Which of the following best describes what happens? Assume temperature is constant.
    a. The pressure in the container doubles.
    b. The pressure in the container increases but does not double.
    c. The pressure in the container more than doubles.
    d. The volume of the container doubles.
    e. The volume of the container more than doubles.

    ANSWER: a.   9.92 L

39. You are holding four balloons each containing 10.0 g of a different gas. The balloon containing which gas is the largest?
    a. H_2
    b. He
    c. Ne
    d. O_2
    e. All have the same volume.

    ANSWER: a. H_2

40. You have two separate containers each filled with a gas. The containers have the same volume and are at the same temperature. The gases also exert the same pressure. Which of the following statements are true?
    a. For conditions of P, V, and T to be the same, the gases must be identical.
    b. For conditions of P, V, and T to be the same, the gases can be different but the number of moles of gas in each balloon must be the same.
    c. For conditions of P, V, and T to be the same, the gases can be different but the mass of gas in each balloon must be the same.
    d. For conditions of P, V, and T to be the same, the gases can be different but the molar mass of gas in each balloon must be the same.
    e. None of the above statements are true.

    ANSWER: b.   For conditions of P, V, and T to be the same, the gases can be different but the number of moles of gas in each balloon must be the same.

41. A specified quantity of an unknown gas has the volume of 14.3 mL at 22°C and 659 torr. Calculate the volume at STP.

    ANSWER: 11.5 mL

42. A 37.4-mL sample of H_2 at STP would contain how many grams of hydrogen?

    ANSWER: 3.37×10^{-3} g

43. A 251-mL sample of a gas at STP is heated to 45°C. The final pressure is 1.10 atm. Calculate the volume of this gas under the new conditions.

ANSWER: 266 mL

44. A gas occupies 15.0 L at STP. What volume will it occupy at 735 torr and 57°C?
    a. 1.2 L
    b. 6.7 L
    c. 9.7 L
    d. 4.6 L
    e. 19 L

ANSWER: e.   19 L

45. A 25.0-L sample of gas at STP is heated to 55°C at 605 torr. What is the new volume?
    a. 38 L
    b. 76 L
    c. 3.5 L
    d. 56 L
    e. 17 L

ANSWER: a.   38 L

Use the following information to answer the next questions:
A gas occupies 30. L at 2.0 atm pressure and 27°C.

46. Calculate its volume if the pressure is decreased to 1.0 atm at constant temperature.
    a. 15 L
    b. 60. L
    c. 120 L
    d. 79 L
    e. 45 L

ANSWER: b.   60. L

47. Calculate its volume if the pressure remains at 2.0 atm, but the temperature is raised to 54°C.
    a. 44 L
    b. 86 L
    c. 12 L
    d. 33 L
    e. 51 L

ANSWER: d.   33 L

48. Calculate its volume at STP.
    a. 33 L
    b. 29 L
    c. 45 L
    d. 18 L
    e. 55 L

ANSWER: e.   55 L

49. How many moles of gas are present in the sample?
    a. 1.2 mol
    b. 4.8 mol
    c. 2.4 mol
    d. 6.8 mol
    e. 9.2 mol

    ANSWER: c.   2.4 mol

50. An oxygen sample has a volume of 4.50 L at 27°C and 800.0 torr. How many oxygen molecules does it contain?
    a. 1.16×10^{23}
    b. 5.8×10^{22}
    c. 2.32×10^{24}
    d. 1.16×10^{22}
    e. none of these

    ANSWER: a. 1.16×10^{23}

51. What volume will 28.0 g of N_2 occupy at STP?
    a. 5.60 L
    b. 11.2 L
    c. 22.4 L
    d. 44.8 L
    e. none of these

    ANSWER: c.   22.4 L

52. A 6.35-L sample of carbon monoxide is collected at 55°C and 0.892 atm. What volume will the gas occupy at 1.05 atm and 20. °C?
    a. 1.96 L
    b. 5.46 L
    c. 4.82 L
    d. 6.10 L
    e. none of these

    ANSWER: c.   4.82 L

53. The volume of a sample of gas is 650. mL at STP. What volume will the sample occupy at 0.0°C and 950. torr?
    a. 476 mL
    b. 520. mL
    c. 568 mL
    d. 650. mL
    e. none of these

    ANSWER: b.   520. mL

180 *Chapter 12*

54. Mercury vapor contains Hg atoms. What is the volume of 200. g of mercury vapor at 822 K and 0.500 atm?
    a. 135 L
    b. 82.2 L
    c. 329 L
    d. 67.2 L
    e. none of these

    ANSWER: a.   135 L

55. What volume is occupied by 19.6 g of methane, CH_4, at 27°C and 1.59 atm?
    a. 1.71 L
    b. 18.9 L
    c. 27.7 L
    d. 302 L
    e. not enough data to calculate

    ANSWER: b.   18.9 L

56. A 4.40-g piece of solid CO_2 (dry ice) is allowed to vaporize (change to $CO_2(g)$) in a balloon. The final volume of the balloon is 1.00 L at 300. K. What is the pressure of the gas?
    a. 2.46 atm
    b. 246 atm
    c. 0.122 atm
    d. 122 atm
    e. none of these

    ANSWER: a.   2.46 atm

Zinc metal is added to hydrochloric acid to generate hydrogen gas and is collected over a liquid whose vapor pressure is the same as pure water at 20.0°C (18 torr). The volume of the mixture is 1.7 L, and its total pressure is 0.810 atm.

57. Determine the partial pressure of the hydrogen gas in this mixture.
    a. 562 torr
    b. 580 torr
    c. 598 torr
    d. 616 torr
    e. 634 torr

    ANSWER: c.   598 torr

58. Determine the number of moles of hydrogen gas present in the sample.
    a. 42 mol
    b. 0.82 mol
    c. 1.3 mol
    d. 0.056 mol
    e. 22 mol

    ANSWER: d.   0.056 mol

59.	A vessel with an internal volume of 10.0 L contains 2.80 g of nitrogen gas, 0.403 g of hydrogen gas, and 79.9 g of argon gas. At 25°C, what is the pressure (in atm) inside the vessel?
	a.	0.471 atm
	b.	6.43 atm
	c.	3.20 atm
	d.	5.62 atm
	e.	2.38 atm

	ANSWER:	d.	5.62 atm

60.	A gas is collected over water at a certain temperature. The total pressure is 762 torr. The vapor pressure of water at this temperature is 17 torr. The partial pressure of the gas collected is
	a.	762 torr
	b.	17 torr
	c.	779 torr
	d.	745 torr
	e.	none of these

	ANSWER:	d.	745 torr

61.	What would happen to the average kinetic energy of the molecules of a gas sample if the temperature of the sample increased from 20°C to 40°C?
	a.	It would double.
	b.	It would increase.
	c.	It would decrease.
	d.	It would become half its value.
	e.	Two of these

	ANSWER:	b.	It would increase.

62.	Which of the following statements is *true* about the kinetic molecular theory?
	a.	The volume of a gas particle is considered to be small--about 0.10 mL.
	b.	Pressure is due to the collisions of the gas particles with the walls of the container.
	c.	Gas particles repel each other but do not attract one another.
	d.	Adding an ideal gas to a closed container will cause an increase in temperature.
	e.	At least two of these statements are correct.

	ANSWER:	b.	Pressure is due to the collisions of the gas particles with the walls of the container.

63.	Which of the following statements is *true* concerning ideal gases?
	a.	The temperature of the gas sample is directly related to the average velocity of the gas particles.
	b.	At STP, 1.0 L of Ar(*g*) contains about twice the number of atoms as 1.0 L of Ne(*g*) because the molar mass of Ar is about twice that of Ne.
	c.	A gas exerts pressure as a result of the collisions of the gas molecules with the walls of the container.
	d.	The gas particles in a sample exert attraction for one another.
	e.	All of the above are false.

	ANSWER:	c.	A gas exerts pressure as a result of the collisions of the gas molecules with the walls of the container.

64. Which conditions of P and T are most ideal for a gas?
    a.   high P, high T
    b.   high P, low T
    c.   low P, high T
    d.   low P, low T
    e.   depends on the gas

    ANSWER: c.   low P, high T

65. Use the kinetic molecular theory of gases to predict what would happen to a closed sample of a gas whose temperature increased while its volume decreased.
    a.   Its pressure would decrease.
    b.   Its pressure would increase.
    c.   Its pressure would hold constant.
    d.   The number of moles of the gas would decrease.
    e.   The average kinetic energy of the molecules of the gas would decrease.

    ANSWER: b.   Its pressure would increase.

66. When acetylene gas, C_2H_2, burns, how many liters of O_2 at STP are used for every 4.0 mol of acetylene burned? (*Hint:* Write the balanced equation for the reaction first.)

    ANSWER: 224 L

67. What volume of HCl(g) measured at STP can be produced from 4.00 g of H_2 and excess Cl_2 according to the following equation:
    $H_2(g) + Cl_2(g) \rightarrow 2HCl(g)$

    ANSWER: 88.9 L

68. How many liters of HCl(g) measured at STP can be produced from 4.00 g of Cl_2 and excess H_2 according to the following equation:
    $H_2(g) + Cl_2(g) \rightarrow 2HCl(g)$

    ANSWER: 2.53 L

69. How many moles of $O_2(g)$ are needed to react completely with 52.0 L of $CH_4(g)$ at STP to produce $CO_2(g)$ and $H_2O(g)$ according to the following reaction:
    $CH_4(g) + 2O_2 \rightarrow CO_2(g) + 2H_2O(g)$
    a.   11.6
    b.   52.0
    c.   4.64
    d.   2.32
    e.   none of these

    ANSWER: c.   4.64

70. Hydrogen peroxide decomposes to form water and oxygen gas according to the following equation:

$$2H_2O_2(aq) \rightarrow + 2H_2O(l) + O_2(g)$$

Suppose 100.0 g of hydrogen peroxide decomposes and all of the oxygen gas is collected in a balloon at 1.00 atm and 25°C. Determine the volume of the balloon.

a. 22.4 L
b. 24.5 L
c. 35.9 L
d. 71.9 L
e. none of these

ANSWER: c.   35.9 L

71. You carry out the reaction represented by the following equation

$$N_2(g) + 3H_2(g) \rightarrow 2NH_3(g)$$

You add an equal number of moles of nitrogen gas and hydrogen gas in a balloon. The volume of the balloon is 1.00 L before any reaction occurs. Determine the volume of the balloon after the reaction is complete.

a. 0.333 L
b. 0.667
c. 1.00 L
d. 1.50 L
e. 3.00 L

ANSWER: b.   0.667 L

72. 2.00 L of NO(g) at 1.00 atm and 25°C is reacted with 1.00 L of $H_2(g)$ at 2.00 atm and 25°C according to the unbalanced equation.

$$NO(g) + H_2(g) \rightarrow N_2(g) + H_2O(g)$$

Which, if either, is the limiting reactant?

ANSWER: There is no limiting reactant.

73. You place 15.0 grams of nitrogen gas and 15.0 grams of hydrogen gas in a container fitted with a massless, frictionless piston. If the original volume of the container is 10.0 L, what is the volume after the reaction has run to completion? Assume constant temperature.

$$N_2(g) + 3H_2(g) \rightarrow 2NH_3(g)$$

ANSWER: 8.66 L

CHAPTER 13
Liquids and Solids

1. At 1 atm of pressure and a temperature of 0°C, which phase(s) of H_2O can exist?
   a. ice and water
   b. ice and water vapor
   c. water only
   d. water vapor only
   e. ice only

   ANSWER: a.   ice and water

2. The normal freezing point of water is
   a. 0°F
   b. 273 K
   c. 32°C
   d. 373°C
   e. none of these

   ANSWER: b.   273 K

3. The normal boiling point of water is
   a. 0°F
   b. 32°F
   c. 273 K
   d. 373 K
   e. none of these

   ANSWER: d.   373 K

4. In general, the density of a compound as a gas is closer in value to that of the compound as a liquid than the density of the compound as a liquid is closer in value to that of the compound as a solid. T/F _____

   ANSWER: False

5. Second-row hydrides generally have higher than expected boiling points for their position on the periodic table. T/F _____

   ANSWER: True

6. More than two-thirds of the earth's surface is covered by water. T/F _____

   ANSWER: True

7. The bonds between hydrogen and oxygen in a water molecule can be characterized as _____.
   a. hydrogen bonds
   b. London forces
   c. intermolecular forces
   d. intramolecular forces
   e. dispersion forces

   ANSWER: d.   intramolecular forces

8. 12,500 J of energy is added to 2.0 mol (36 g) of H_2O as an ice sample at 0°C. The mole heat of fusion is 6.02 kJ/mol. The specific heat of liquid water is 4.18 J/mol K. The molar heat of vaporization is 40.6 kJ/mol. The resulting sample contains which of the following?
   a. only ice
   b. ice and water
   c. only water
   d. water and water vapor
   e. only water vapor

   ANSWER: c.   only water

9. Calculate the quantity of energy required to change 3.00 mol of liquid water to steam at 100°C. The molar heat of vaporization of water is 40.6 kJ/mol.
   a. 40.6 kJ
   b. 13.5 kJ
   c. 122 kJ
   d. 300 kJ
   e. none of these

   ANSWER: c.   122 kJ

10. Calculate the quantity of energy required to change 26.5 g of liquid water to steam at 100°C. The molar heat of vaporization of water is 40.6 kJ/mol.
    a. 1.08×10^3 kJ
    b. 59.8 kJ
    c. 1.53 kJ
    d. 27.6 kJ
    e. none of these

    ANSWER: b.   59.8 kJ

11. The specific heat capacity of liquid water is 4.18 kJ/g °C. Calculate the quantity of energy required to heat 1.00 g of water from 26.5°C to 83.7°C.
    a. 239 J
    b. 57.2 J
    c. 13.7 J
    d. 350. J
    e. none of these

    ANSWER: a.   239 J

12. The specific heat capacity of liquid water is 4.18 J/g °C. Calculate the quantity of energy required to heat 10.0 g of water from 26.5°C to 83.7°C.
    a. 837 J
    b. 572 J
    c. 239 J
    d. 2.39×10^3 J
    e. none of these

    ANSWER: d. 2.39×10^3 J

13. The molar heat of fusion of water is 6.02 kJ/mol. Calculate the energy required to melt 46.8 g of water.
    a. 282 kJ
    b. 2.32 kJ
    c. 6.02 kJ
    d. 7.77 kJ
    e. none of these

    ANSWER: e.   none of these

14. The molar heat of fusion of water is 6.02 kJ/mol. Calculate the energy required to melt 3.00 mol of ice.
    a. 6.02 kJ/mol
    b. 12.0 kJ/mol
    c. 18.1 kJ/mol
    d. 2.01 kJ/mol
    e. none of these

    ANSWER: c.   18.1 kJ/mol

15. The molar heat of fusion of water is 6.02 kJ/mol. Calculate the energy required to melt 14.0 g of water.
    a. 6.02 kJ
    b. 4.68 kJ
    c. 84.3 kJ
    d. 7.74 kJ
    e. none of these

    ANSWER: b.   4.68 kJ

16. The molar heats of fusion of water and iodine are 6.02 kJ/mol and 16.7 kJ/mol, respectively. It will take more energy to melt a gram of ice than to melt a gram of solid iodine. T/F

     ————

     ANSWER: False

17. When a water molecule forms a hydrogen bond with another water molecule, which atoms are involved in the interaction?
     a. a hydrogen from one molecule and a hydrogen from the other molecule
     b. a hydrogen from one molecule and an oxygen from the other molecule
     c. an oxygen from one molecule and an oxygen from the other molecule
     d. two hydrogens from one molecule and one oxygen from the other molecule
     e. two hydrogens from one molecule and one hydrogen from the other molecule

     ANSWER: b.   a hydrogen from one molecule and an oxygen from the other molecule

18. The freezing point of helium is approximately -270°C. The freezing point of xenon is -112°C. Both of these are in the noble gas family. Which of the following statements is supported by these data?
     a. Helium and xenon form highly polar molecules.
     b. As the molar mass of the noble gas increases, the freezing point decreases.
     c. The London forces between the helium molecules are greater than the London forces between the xenon molecules.
     d. The London forces between the helium molecules are less than the London forces between the xenon molecules.
     e. none of these

     ANSWER: d.   The London forces between the helium molecules are less than the London forces between the xenon molecules.

19. Choose the state of water in which the water molecules are farthest apart on average.
     a. steam (vapor)
     b. liquid
     c. ice (solid)
     d. all the same

     ANSWER: a.   steam (vapor)

20. Which of the following has the lowest vapor pressure?
     a. H_2O
     b. NaCl
     c. NH_3
     d. O_2
     e. CH_4

     ANSWER: b.   NaCl

21. Which of the following should have the lowest boiling point?
    a. CH_4
    b. C_2H_6
    c. C_3H_8
    d. C_4H_{10}
    e. C_5H_{12}

    ANSWER: a. CH_4

22. Order the intermolecular forces (dipole-dipole, London dispersion, ionic, and hydrogen bonding) from weakest to strongest.
    a. dipole-dipole, London dispersion, ionic, hydrogen bonding
    b. London dispersion, dipole-dipole, hydrogen bonding, ionic
    c. hydrogen bonding, dipole-dipole, London dispersion, ionic
    d. dipole-dipole, ionic, London dispersion, hydrogen bonding
    e. London dispersion, ionic, dipole-dipole, hydrogen bonding

    ANSWER: b.   London dispersion, dipole-dipole, hydrogen bonding, ionic

Identify the major attractive force in each of the following molecules.

23. CH_4
    a. dipole-dipole
    b. London dispersion
    c. ionic
    d. hydrogen bonding
    e. none of these

    ANSWER: b.   London dispersion

24. O_2
    a. dipole-dipole
    b. London dispersion
    c. ionic
    d. hydrogen bonding
    e. none of these

    ANSWER: b.   London dispersion

25. CO
    a. dipole-dipole
    b. London dispersion
    c. ionic
    d. hydrogen bonding
    e. none of these

    ANSWER: a.   dipole-dipole

26. K_2O
    a. dipole-dipole
    b. London dispersion
    c. ionic
    d. hydrogen bonding
    e. none of these

    ANSWER: c.   ionic

27. H_2O
    a. dipole-dipole
    b. London dispersion
    c. ionic
    d. hydrogen bonding
    e. none of these

    ANSWER: d.   hydrogen bonding

28. The intermolecular forces called hydrogen bonding will not exist between molecules of
    a. H_2O
    b. H_2
    c. NH_3
    d. HF
    e. any of these

    ANSWER: b. H_2

29. The process of sublimation happens when which of the following occurs?
    a. A solid becomes a liquid.
    b. A liquid becomes a solid.
    c. A liquid becomes a gas.
    d. A gas becomes a liquid.
    e. A solid becomes a gas.

    ANSWER: e.   A solid becomes a gas.

30. Which of the following processes must exist in equilibrium with the evaporation process when a measurement of vapor pressure is made?
    a. fusion
    b. vaporization
    c. sublimation
    d. boiling
    e. condensation

    ANSWER: e.   condensation

31. Water sits in an open beaker. Assuming constant temperature and pressure, the vapor pressure of the water _____ as the water evaporates.
    a. increases
    b. decreases
    c. stays the same

    ANSWER: c.   stays the same

32. The normal boiling point of liquid X is less than that of Y, which is less than that of Z. Which of the following is the correct order of increasing vapor pressure of the three liquids at STP?
    a. X, Y, Z
    b. Z, Y, X
    c. Y, X, Z
    d. X, Z, Y
    e. Y, Z, X

    ANSWER: b.   Z, Y, X

33. The P_{vap} for water at 100.0°C is
    a. 85 torr
    b. 760 torr
    c. 175 torr
    d. 1 torr
    e. More information is needed.

    ANSWER: b.   760 torr

34. Of the following substances, choose the one with the greatest vapor pressure.
    a. He*(l)*
    b. Ne*(l)*
    c. Ar*(l)*
    d. Xe*(l)*
    e. Rn*(l)*

    ANSWER: a.   He*(l)*

35. As the temperature of a liquid increases, the vapor pressure of the liquid generally
    a. increases
    b. decreases
    c. stays the same
    d. depends on the type of intermolecular forces

    ANSWER: a.   increases

36. Which of the following is true for ionic solids dissolved in water?
    a. The solution will conduct electricity.
    b. The solid will dissolve into neutral molecules.
    c. The solution will not conduct electricity.
    d. The ions in solution will form a large crystal.
    e. none of these

    ANSWER: a.   The solution will conduct electricity.

37. Which of the following has the highest melting temperature?
    a. H_2O
    b. CO_2
    c. S_8
    d. MgF_2
    e. P_4

    ANSWER: d. MgF_2

38. Name the type of crystalline solid formed by potassium bromide.
    a. molecular solid
    b. atomic solid
    c. ionic solid
    d. amorphous solid
    e. none of these

    ANSWER: c. ionic solid

39. Name the type of crystalline solid formed by copper.
    a. molecular solid
    b. atomic solid
    c. ionic solid
    d. amorphous solid
    e. none of these

    ANSWER: b. atomic solid

40. Name the type of crystalline solid formed by SiO_2.
    a. molecular solid
    b. atomic solid
    c. ionic solid
    d. amorphous solid
    e. none of these

    ANSWER: a. molecular solid

41. An alloy has metallic properties. T/F _____

    ANSWER: True

CHAPTER 14

Solutions

1. One mole of each of the following compounds is added to water in separate flasks to make 1.0 L of solution. Which solution has the largest *total* ion concentration?
    a. calcium carbonate
    b. potassium phosphate
    c. aluminum hydroxide
    d. silver chloride
    e. sodium chloride

    ANSWER: b.   potassium phosphate

2. In soda pop, $CO_2(g)$ is a _____ and water is the _____.
    a. solute; solvent
    b. solvent; solute
    c. solution; solute
    d. solute; solution
    e. solvent; solution

    ANSWER: a.   solute; solvent

3. A mixture of sand and water is a(n) _____.
    a. solution
    b. solvent
    c. solute
    d. aqueous solution
    e. none of these

    ANSWER: e.   none of these

4. Approximately 38 g of NaCl can be dissolved in 100 g of water at 25°C. A solution prepared by adding 35 g of NaCl to 100 g of water at 25°C is unsaturated. T/F _____

    ANSWER: True

5. What term is used by chemists to quantitatively describe a solution in which a relatively small amount of solute is dissolved?
   a. dilute
   b. saturated
   c. supersaturated
   d. concentrated
   e. unsaturated

   ANSWER: a.   dilute

6. When a solvent has dissolved all the solute it can at a particular temperature, it is said to be
   a. diluted
   b. unsaturated
   c. supersaturated
   d. saturated
   e. none of these

   ANSWER: d.   saturated

7. The total mass of a solution is 153.4 g. The solvent mass is 125.2 g. What is the mass percent of the solute?
   a. 81.62%
   b. 122.5%
   c. 18.38%
   d. 22.52%
   e. not enough information given

   ANSWER: c.   18.38%

8. An oven-cleaning solution is 40.0% (by mass) NaOH. If one jar of this product contains 454 g of solution, how much NaOH does it contain?
   a. 114 g
   b. 1.14×10^3 g
   c. 182 g
   d. 18.2 g
   e. none of these

   ANSWER: c.   182 g

9. A nitric acid solution containing 71.0% HNO_3 (by mass) has a density of 1.42 g/mL. How many moles of HNO_3 are present in 1.00 L of this solution?
   a. 1.60 mol
   b. 16.0 mol
   c. 1.13 mol
   d. 11.3 mol
   e. none of these

   ANSWER: b.   16.0 mol

10. A nitric acid solution that is 70.0% HNO_3 (by mass) contains
    a. 70.0 g HNO_3 and 100.0 g water
    b. 70.0 mol HNO_3
    c. 70.0 g HNO_3 and 30.0 g water
    d. 70.0 g HNO_3 and 70.0 g water
    e. none of these

    ANSWER: c.   70.0 g HNO_3 and 30.0 g water

11. A 100.0-g sample of a nitric acid solution that is 70.0% HNO_3 (by mass) contains
    a. 70.0 mol HNO_3
    b. 1.11 mol HNO_3
    c. 3.89 mol HNO_3
    d. 4.41×10^3 mol HNO_3
    e. none of these

    ANSWER: b.   1.11 mol HNO_3

12. You have two solutions of sodium chloride. One is a 2.00 M solution, the other is a 4.00 M solution. You have much more of the 4.00 M solution, and you add the solutions together. Which of the following could be the concentration of the final solution?
    a. 2.60 M
    b. 3.00 M
    c. 3.80 M
    d. 6.00 M
    e. 7.20 M

    ANSWER: c.   3.80 M

13. Determine the concentration of a solution made by dissolving 10.0 g of sodium chloride in 750.0 mL of solution.
    a. 0.0133 M
    b. 0.133 M
    c. 0.171 M
    d. 0.228 M
    e. 0.476 M

    ANSWER: d.   0.228 M

14. If you mix 20.0 mL of a 3.00 M sugar solution with 30.0 mL of a 5.00 M sugar solution, you will end up with a sugar solution of _____.
    a. 2.00 M
    b. 3.80 M
    c. 4.00 M
    d. 4.20 M
    e. 8.00 M

    ANSWER: d.   4.20 M

15. You have 250.0 mL of 4.00 M sugar solution. You add 300.0 mL of water to this solution. Determine the concentration of the sugar solution after the water has been added.
    a. 3.33 *M*
    b. 2.18 *M*
    c. 2.00 *M*
    d. 1.82 *M*
    e. 1.73 *M*

    ANSWER: d.   1.82 *M*

16. Calculate the concentration of chloride ions when 100.0 mL of 0.200 *M* sodium chloride is mixed with 250.0 mL of 0.150 *M* calcium chloride.
    a. 0.095 *M*
    b. 0.164 *M*
    c. 0.271 *M*
    d. 0.350 *M*
    e. none of these

    ANSWER: c.   0.271 *M*

17. Which of the following aqueous solutions contains the greatest number of ions in solution?
    a. 2.0 L of 1.50 *M* sodium phosphate
    b. 2.0 L of 1.50 *M* magnesium chloride
    c. 3.0 L of 1.50 *M* sodium chloride
    d. 2.0 L of 2.00 *M* potassium fluoride
    e. 1.0 L of 3.00 *M* sodium sulfate

    ANSWER: a.   2.0 L of 1.50 *M* sodium phosphate

18. You have 25.00 mL of a 0.1000 *M* sugar solution. How much water must be added to make a 0.025 *M* solution?
    a. 6.250 mL
    b. 10.00 mL
    c. 25.00 mL
    d. 75.00 mL
    e. 100.0 mL

    ANSWER: d.   75.00 mL

19. What is the minimum volume of a 2.00 *M* NaOH solution needed to make 150.0 mL of a 0.800 *M* NaOH solution?
    a. 375 mL
    b. 150. mL
    c. 120. mL
    d. 90.0 mL
    e. 60.0 mL

    ANSWER: e.   60.0 mL

196   *Chapter 14*

20. You have 100.0 mL of a 0.2500 *M* solution of NaCl sitting in a beaker. After several days you test the solution and find that it is now 0.3125 *M*. How much water must have evaporated?
    a. 10.0 mL
    b. 20.0 mL
    c. 80.0 mL
    d. 90.0 mL
    e. The concentration will not change because of evaporation.

    ANSWER: b. 20.0 mL

21. A solution is prepared by dissolving 7.31 g of Na_2SO_4 in enough water to make 225 mL of solution. Calculate the solution molarity.
    a. 30.8 *M*
    b. 1.64 *M*
    c. 0.136 *M*
    d. 0.229 *M*
    e. 3.11 *M*

    ANSWER: d. 0.229 *M*

22. What volume of a 0.209 *M* Na_2S solution contains 1.01 g of Na^+ ions?
    a. 0.21 L
    b. 0.651 L
    c. 3.78 L
    d. 1.43 L
    e. 0.105 L

    ANSWER: e. 0.105 L

23. How many grams of $CaCl_2$ (molar mass = 111.0 g/mol) are needed to prepare 5.91 L of 0.500 *M* $CaCl_2$ solution?
    a. 55.5 g
    b. 328 g
    c. 111 g
    d. 297 g
    e. 198 g

    ANSWER: b. 328 g

24. A 0.251-g sample of NaCl (molar mass = 58.44 g/mol) is dissolved in enough water to make 5.20 mL of solution. Calculate the molarity of the resulting solution.
    a. 0.826 *M*
    b. 1.65 *M*
    c. 0.413 *M*
    d. 0.756 *M*
    e. 2.66 *M*

    ANSWER: a. 0.826 *M*

25. What mass of calcium chloride, $CaCl_2$ is needed to prepare 2.850 L of a 1.56 M solution?
    a. 25.9 g
    b. 60.8 g
    c. 111 g
    d. 203 g
    e. 493 g

    ANSWER: e.   493 g

26. A 54.8-g sample of $SrCl_2$ is dissolved in 112.5 mL of solution. Calculate the molarity of this solution.
    a. 0.346 M
    b. 3.07 M
    c. 3.96 M
    d. 8.89 M
    e. none of these

    ANSWER: b.   3.07 M

27. A 75.80-g sample of NaCl is dissolved in 250.0 mL of solution. Calculate the molarity of this solution.
    a. 2.370 M
    b. 4.000 M
    c. 5.188 M
    d. 233.8 M
    e. none of these

    ANSWER: c.   5.188 M

28. Calculate the mass of silver nitrate (in grams) in a 145 mL solution of 4.31 M $AgNO_3$.
    a. 5.05 g
    b. 24.6 g
    c. 106 g
    d. 170. g
    e. none of these

    ANSWER: c.   106 g

29. Calculate the molarity of a solution prepared by dissolving 4.09 g of NaI in enough water to prepare 312 mL of solution.
    a. 76.3 M
    b. 0.0131 M
    c. 0.0875 M
    d. 0.175 M
    e. 1.66 M

    ANSWER: c.   0.875 M

30. A 20.0-g sample of HF is dissolved in enough water to give 2.0 x 10^2 mL of solution. The concentration of the solution is
    a.   1.0 *M*
    b.   3.0 *M*
    c.   0.10 *M*
    d.   5.0 *M*
    e.   10.0 *M*

    ANSWER: d.   5.0 *M*

31. How many grams of NaCl are contained in 350. mL of a 0.250 *M* solution of sodium chloride?
    a.   41.7 g
    b.   5.11 g
    c.   14.6 g
    d.   87.5 g
    e.   none of these

    ANSWER: b.   5.11 g

32. The molarity of Cl⁻ in 110. mL of a solution containing 5.55 g of $CaCl_2$ is
    a.   0.00550 *M*
    b.   0.455 *M*
    c.   0.668 *M*
    d.   0.909 *M*
    e.   50.5 *M*

    ANSWER: d.   0.909 *M*

33. A 12.6-g sample of potassium sulfate is dissolved in 350.0 mL of solution. Calculate the molarity of this solution.
    a.   2.86 *M*
    b.   0.498 *M*
    c.   0.266 *M*
    d.   0.207 *M*
    e.   none of these

    ANSWER: d.   0.207 *M*

34. To make 44.6 mL of a 0.405 *M* solution of NaCl, what mass of NaCl is required?
    a.   1.08 g
    b.   2.66 g
    c.   23.7 g
    d.   58.4 g
    e.   none of these

    ANSWER: a.   1.08 g

35. What volume of a 0.550 *M* solution of potassium hydroxide can be made with 25.0 g of potassium hydroxide?
    a. 1234 mL
    b. 810. mL
    c. 550. mL
    d. 450. mL
    e. none of these

    ANSWER: b.   810. mL

36. What mass of solute is contained in 256 mL of a 0.895 *M* ammonium chloride solution?
    a. 12.3 g
    b. 13.7 g
    c. 47.9 g
    d. 53.5 g
    e. none of these

    ANSWER: a.   12.3 g

37. What mass of solute is contained in 417 mL of a 0.314 *M* magnesium fluoride solution?
    a. 8.15 g
    b. 19.6 g
    c. 26.0 g
    d. 62.3 g
    e. none of these

    ANSWER: a.   8.15 g

38. What mass of solute is contained in 25.4 mL of a 1.56 *M* potassium bromide solution?
    a. 3.02 g
    b. 4.72 g
    c. 119 g
    d. 186 g
    e. none of these

    ANSWER: b.   4.72 g

39. A 50.0-g sample of NaCl is added to water to create a 0.50 L solution. What volume of this solution is needed to create a 1.30 L solution that is 0.100 *M* NaCl?
    a. 152 mL
    b. 255 mL
    c. 87.6 mL
    d. 412 mL
    e. 10.1 mL

    ANSWER: a.   152 mL

40. A chemist needs 225 mL of 2.4 M HCl. What volume of 12 M HCl must be dissolved in water to form this solution?
    a. 3.4 mL
    b. 7.2 mL
    c. 21 mL
    d. 6.8 mL
    e. 45 mL

    ANSWER: e.   45 mL

41. What volume of 18.0 M sulfuric acid is required to prepare 16.5 L of 0.126 M H_2SO_4?
    a. 11.6 mL
    b. 0.116 L
    c. 232 mL
    d. 0.264 L
    e. 1.16 L

    ANSWER: b.   0.116 L

42. What volume of 12.0 M nitric acid is required to prepare 6.00 L of 0.100 M nitric acid?
    a. 1.20 L
    b. 1.00 L
    c. 0.500 L
    d. 0.0500 L
    e. 0.0200 L

    ANSWER: d.   0.0500 L

43. A 51.24-g sample of $Ba(OH)_2$ is dissolved in enough water to make 1.20 L of solution. How many milliliters of this solution must be diluted with water in order to make 1.00 L of 0.100 M $Ba(OH)_2$?
    a. 400. mL
    b. 333 mL
    c. 278 mL
    d. 1.20×10^3 mL
    e. none of these

    ANSWER: a.   400. mL

44. What volume of 18.0 M sulfuric acid must be used to prepare 15.5 L of 0.195 M H_2SO_4?
    a. 168 mL
    b. 0.336 L
    c. 92.3 mL
    d. 226 mL
    e. none of these

    ANSWER: a.   168 mL

45. What volume of 18.0 *M* H_2SO_4 is required to prepare 12.0 L of 0.156 *M* sulfuric acid?
    a. 1.87 mL
    b. 52.0 mL
    c. 208 mL
    d. 104 mL
    e. 1.04 L

    ANSWER: d. 104 mL

You have 3.00 L of a 3.00 *M* solution of NaCl*(aq)* called solution A. You also have 2.00 L of a 2.00 *M* solution of $AgNO_3$*(aq)* called solution B. You mix these solutions together, making solution C.

Calculate the concentrations (in *M*) of the following ions in solution C.

46. Na^+
    a. 0 *M*
    b. 0.800 *M*
    c. 1.00 *M*
    d. 1.50 *M*
    e. 1.80 *M*

    ANSWER: e. 1.80 *M*

47. Cl^-
    a. 0 *M*
    b. 0.800 *M*
    c. 1.00 *M*
    d. 1.50 *M*
    e. 1.80 *M*

    ANSWER: c. 1.00 *M*

48. Ag^+
    a. 0 *M*
    b. 0.800 *M*
    c. 1.00 *M*
    d. 1.50 *M*
    e. 1.80 *M*

    ANSWER: a. 0 *M*

49. NO_3^-
    a. 0 *M*
    b. 0.800 *M*
    c. 1.00 *M*
    d. 1.50 *M*
    e. 1.80 *M*

    ANSWER: b. 0.800 *M*

50. You react 250.0 mL of a 0.10 M barium nitrate solution with 200.0 mL of a 0.10 M potassium phosphate solution. What ions are left in solution after the reaction is complete?
    a. potassium ion, nitrate ion, barium ion
    b. potassium ion, nitrate ion
    c. potassium ion, barium ion, phosphate ion
    d. barium ion, nitrate ion, phosphate ion
    e. potassium ion, nitrate ion, phosphate ion

    ANSWER: e.   potassium ion, nitrate ion, phosphate ion

51. Magnesium metal reacts with hydrochloric acid to form magnesium chloride and hydrogen gas. Suppose we react an excess of magnesium metal with 20.0 mL of a 3.00 M solution of hydrochloric acid and collect all of the hydrogen in a balloon at 25°C and 1.00 atm. What is the expected volume of the balloon?
    a. 0.672 L
    b. 0.734 L
    c. 1.34 L
    d. 1.47 L
    e. 22.4 L

    ANSWER: b.   0.734 L

52. If you mix 40.0 mL of a 0.200 M solution of K_2CrO_4 is reacted with 40.0 mL of a 0.200 M solution of $AgNO_3$, what mass of solid forms?
    a. 0.896 g
    b. 1.33 g
    c. 1.79 g
    d. 2.65 g
    e. none of these

    ANSWER: b.   1.33 g

53. What is the limiting reactant when 15.0 mL of 0.150 M lead(II) nitrate solution is reacted with 20.0 mL of 0.200 M sodium iodide solution?
    a. $PbNO_3$
    b. $Pb(NO_3)_2$
    c. NaI
    d. PbI_2
    e. There is no limiting reactant.

    ANSWER: c.   NaI

54. Consider the reaction between 50.0 mL of 0.200 M NaOH solution and 75.0 mL of a 0.100 M HCl solution. Which of the following statements is true?
    a. After the reaction, the concentration of Na^+ is greater than the concentration of the OH^-.
    b. The NaOH is the limiting reactant.
    c. After the reaction, the concentration of Na^+ is equal to the concentration of Cl^+.
    d. After the reaction, the concentration of Na^+ is still 0.200 M because Na^+ is a spectator ion.
    e. none of these

    ANSWER: a.   After the reaction, the concentration of Na^+ is greater than the concentration of the OH^-.

Copyright © Houghton Mifflin Company. All rights reserved.

55. You mix 100.0 mL of a 0.100 *M* NaOH solution and 150.0 mL of a 0.100 *M* HCl solution. Determine the concentration of H^+ in the final mixture after the reaction is complete.
    a. 0.150 *M*
    b. 0.100 *M*
    c. 0.0600 *M*
    d. 0.0500 *M*
    e. 0.0200 *M*

    ANSWER: e.   0.0200 *M*

56. Calculate the volume of 0.125 *M* HNO_3 required to neutralize 25.0 mL of 0.250 *M* NaOH.
    a. 25.0 mL
    b. 50.0 mL
    c. 12.5 mL
    d. 75.0 mL
    e. none of these

    ANSWER: b.   50.0 mL

57. Calculate the volume of 0.500 *M* KOH required to neutralize 150 mL of 0.100 *M* HCl.
    a. 150 mL
    b. 30. mL
    c. 750 mL
    d. 75 mL
    e. none of these

    ANSWER: b.   30. mL

58. In the following acid-base neutralization, 1.68 g of the solid acid HC_6H_5O neutralized 11.61 mL of aqueous NaOH solution base by the reaction
    $NaOH(aq) + HC_6H_5O(aq) \rightarrow H_2O(l) + NaC_6H_5O(aq)$
    Calculate the molarity of the base solution.

    ANSWER: 1.54 *M*

59. Assume that vinegar is a 0.852 *M* solution of acetic acid $(HC_2H_3O_2)$ in water. What volume of 0.2136 *M* NaOH would be needed to completely neutralize 5.26 mL of vinegar?

    ANSWER: 21.0 mL

60. What is the normality of 1.00 L of a sodium hydroxide solution that contains 20.0 g of NaOH?
    a. 2.00 *N*
    b. 1.00 *N*
    c. 1.50 *N*
    d. 0.50 *N*
    e. none of these

    ANSWER: d.   0.50 *N*

61. Determine the normality of a base if 93.0 mL of 1.5 *M* HCl is required to neutralize 52.0 mL of it.
     a. 1.3 *N*
     b. 6.4 *N*
     c. 9.1 *N*
     d. 5.5 *N*
     e. 2.7 *N*

ANSWER: e.   2.7 *N*

CHAPTER 15
Acids and Bases

---

1. Consider the reaction $HNO_2(aq) + H_2O(l) \rightarrow H_3O^+(aq) + NO_2^-(aq)$. Which species is the conjugate base?
   a. $HNO_2(aq)$
   b. $H_2O(l)$
   c. $H_3O^+(aq)$
   d. $NO_2^-(aq)$
   e. two of these

   ANSWER: d. $NO_2^-(aq)$

2. Identify the Bronsted acids and bases in the following equation (a = Bronsted acid, B = Bronsted base):

   $$HSO_3^- + CN^- \rightarrow HCN + SO_3^{2-}$$

a.	B	A	B	A
b.	B	B	A	A
c.	A	B	A	B
d.	A	B	B	A
e.	B	A	A	B

   ANSWER: c. A B A B

3. According to the Bronsted-Lowry definition, a base is
   a. a substance that increases the hydroxide ion concentration in water
   b. a substance that can accept a proton from an acid
   c. a substance that can donate an electron pair to the formation of a covalent bond
   d. a substance that increases the anion formed by the autoionization of the solvent
   e. none of these

   ANSWER: b. a substance that can accept a proton from an acid

4. Which of the following is a conjugate acid-base pair?
    a. HCl/OCl^-
    b. H_2SO_4/SO_4^{2-}
    c. NH_4^+/NH_3
    d. H_3O^+/OH^-
    e. none of these

    ANSWER: c. NH_4^+/NH_3

5. Choose the case that is *not* a conjugate acid-base pair.
    a. HCO_3^-, CO_3^{2-}
    b. H_3O^+, H_2O
    c. OH^-, O^{2-}
    d. H_3PO_4, HPO_4^{2-}
    e. $NH_2OH_2^+, NH_2OH$

    ANSWER: d. H_3PO_4, HPO_4^{2-}

6. Choose the case that is not a Bronsted conjugate acid-base pair.
    a. NH_3, NH_4^+
    b. $C_2O_4^{2-}, HC_2O_4^-$
    c. $HC_2H_3O_2, H_2C_2H_3O_2^+$
    d. O^{2-}, OH^+
    e. SO_3^{2-}, SO_4^{2-}

    ANSWER: e. SO_3^{2-}, SO_4^{2-}

7. The conjugate base of a weak acid is
    a. a strong base
    b. a weak base
    c. a strong acid
    d. a weak acid
    e. none of these

    ANSWER: b.   a weak base

8. In deciding which of two acids is the stronger, one must know
    a. the concentration of each acid solution
    b. the pH of each acid solution
    c. the equilibrium constant of each acid
    d. all of the above
    e. both a and c

    ANSWER: c.   the equilibrium constant of each acid

9. The fact that HCl(*aq*) is a strong acid also means that Cl⁻ is a(n) _____.
   a. strong conjugate base
   b. weak conjugate base
   c. weak acid
   d. amphoteric substance
   e. proton donor

   ANSWER: b.   weak conjugate base

10. Which of the following is the strongest conjugate base?
    a. Cl⁻
    b. NO₃⁻
    c. HSO₄⁻
    d. C₂H₃O₂⁻ or CH₃COO⁻
    e. all the same

    ANSWER: d.   C₂H₃O₂⁻ or CH₃COO⁻

11. Which of the following is *not* a strong acid?
    a. HCl
    b. HNO₃
    c. H₂SO₄
    d. HC₂H₃O₂ or CH₃COOH
    e. All are strong acids.

    ANSWER: d.   HC₂H₃O₂ or CH₃COOH

12. Which of the following must be *true* if a solution is to be considered acidic?
    a. [H⁺] > [OH⁻]
    b. [H⁺] < [OH⁻]
    c. [H⁺] = [OH⁻]
    d. K_w = [H⁺]/[OH⁻]
    e. two of these

    ANSWER: a.   [H⁺] > [OH⁻]

13. As water is heated, its [H⁺] increases. This means that
    a. the water is no longer neutral
    b. [H⁺] > [OH⁻]
    c. [OH⁻] > [H⁺]
    d. a and b are correct.
    e. none of these

    ANSWER: e.   none of these

14. Which has the higher [H^+], 0.20 M NaOH or 0.10 M Ca(OH)$_2$?
    a. 0.20 M NaOH
    b. 0.10 M Ca(OH)$_2$
    c. The [H^+]'s are equal.
    d. The answer depends on the volumes of the respective solutions.

    ANSWER: c.   The [H^+]'s are equal.

15. A substance like water that behaves as an acid or base is said to be amphibasic. T/F
    _____

    ANSWER: False

16. Choose the pair of concentrations that cannot be in a given aqueous solution at 25°C.
    a. [H^+] = 10^{-3} M, [OH^-] = 10^{-11} M
    b. [H^+] = 10^{-7} M, [OH^-] = 10^{-7} M
    c. [H^+] = 10^{-13} M, [OH^-] = 1 M
    d. [H^+] = 10 M, [OH^-] = 10^{-15} M
    e. All of these can exist.

    ANSWER: e.   All of these can exist.

17. A solution where [H^+] = 10^{-13} M is _____.
    a. basic
    b. neutral
    c. acidic
    d. strongly acidic
    e. two of these

    ANSWER: a.   basic

18. A solution has [H^+] = 4.2 x 10^{-3} M. The [OH^-] in this solution is
    a. 4.2 x 10^{-3} M
    b. 4.2 x 10^{-11} M
    c. 2.4 x 10^{-12} M
    d. 1.0 x 10^{-14} M
    e. none of these

    ANSWER: c.   2.4 x 10^{-12} M

19. A solution has [OH^-] = 2.8 x 10^{-7} M. The [H^+] in this solution is
    a. 1.0 M
    b. 2.8 x 10^{-7} M
    c. 1.0 x 10^{-7} M
    d. 3.6 x 10^{-8} M
    e. none of these

    ANSWER: d.   3.6 x 10^{-8} M

20. Calculate the [H$^+$] in a solution that has a pH of 8.95.
    a. $1.1 \times 10^{-2}\ M$
    b. $8.9 \times 10^{-6}\ M$
    c. $1.0 \times 10^{-7}\ M$
    d. $1.1 \times 10^{-9}\ M$
    e. none of these

    ANSWER: d. $1.1 \times 10^{-9}\ M$

21. Calculate the [OH$^-$] in a solution that has a pH of 3.70.
    a. $5.0 \times 10^{-11}\ M$
    b. $1.0 \times 10^{-7}\ M$
    c. $5.0 \times 10^{-4}\ M$
    d. $5.0 \times 10^{-1}\ M$
    e. none of these

    ANSWER: a. $5.0 \times 10^{-11}\ M$

22. A solution with a pH of 2 is how many times more acidic as a solution with a pH of 4?
    a. 0.5
    b. 2
    c. 10
    d. 100
    e. 1000

    ANSWER: d. 100

23. A solution with a pH of 3 is how many times as acidic as a solution with a pH of 4?
    a. 100 times as acidic
    b. 80 times as acidic
    c. 10 times as acidic
    d. 15 times as acidic
    e. 12 times as acidic

    ANSWER: c.   10 times as acidic

24. Calculate the [H$^+$] in a solution that shows a pH of 11.70.
    a. $2.3\ M$
    b. $11.7\ M$
    c. $5.0 \times 10^{-3}\ M$
    d. $2.0 \times 10^{-12}\ M$
    e. none of these

    ANSWER: d. $2.0 \times 10^{-12}\ M$

25. Calculate the [H$^+$] in a solution that shows a pH of 2.30.
    a. 2.3 M
    b. 11.7 M
    c. 5.0 x 10^{-3} M
    d. 2.0 x 10^{-12} M
    e. none of these

    ANSWER: c.   5.0 x 10^{-3} M

26. The pH of a solution at 25°C in which [OH$^-$] = 3.4 x 10^{-5} M is
    a. 4.47
    b. 10.47
    c. 9.53
    d. 6.34
    e. none of these

    ANSWER: c.   9.53

27. Solid calcium hydroxide is dissolved in water until the pH of the solution is 10.94. The hydroxide ion concentration [OH$^-$] of the solution is
    a. 1.1 x 10^{-11} M
    b. 3.06 M
    c. 8.7 x 10^{-4} M
    d. 1.0 x 10^{-13} M
    e. none of these

    ANSWER: c.   8.7 x 10^{-4} M

28. A solution has [H$\cdot$] = 4.0 x 10^{-8} M. The pH of this solution is
    a. 3.20
    b. 6.60
    c. 7.40
    d. 10.80
    e. none of these

    ANSWER: c.   7.40

29. A solution has [H$^+$] = 4.0 x 10^{-8} M. The pOH of this solution is
    a. 3.20
    b. 6.60
    c. 7.40
    d. 10.80
    e. none of these

    ANSWER: b.   6.60

30. Calculate the [H$^+$] in a solution that has a pH of 5.21.
    a. $1.6 \times 10^{-9}\ M$
    b. $6.2 \times 10^{-6}\ M$
    c. $4.0 \times 10^{-3}\ M$
    d. $1.6 \times 10^{-2}\ M$
    e. none of these

    ANSWER: b. $6.2 \times 10^{-6}\ M$

31. A solution has a pH of 3.45. The pOH of this solution is
    a. 3.45
    b. 3.55
    c. 10.45
    d. 10.55
    e. none of these

    ANSWER: d.   10.55

32. A solution has [OH$^-$] = $5.0 \times 10^{-4}\ M$. The pH of this solution is
    a. 3.30
    b. 2.0×10^{-11}
    c. 4.50
    d. 10.70
    e. none of these

    ANSWER: d.   10.70

33. A solution has a pH of 4.35. The [H$^+$] in this solution is
    a. $4.35 \times 10^{-5}\ M$
    b. $4.50 \times 10^{-5}\ M$
    c. $4.35\ M$
    d. $3.50 \times 10^{-4}\ M$
    e. none of these

    ANSWER: b. $4.50 \times 10^{-5}\ M$

34. What is the pH of a solution that has [OH$^-$] = $4.0 \times 10^{-9}\ M$?
    a. 8.40
    b. 5.60
    c. 9.40
    d. 4.60
    e. none of these

    ANSWER: b.   5.60

35. What is the pH of a solution prepared by dissolving 120.0 g NaOH in enough water to make 3.750 L of solution?
    a. 1.03
    b. 9.34
    c. 12.97
    d. 15.03
    e. none of these

    ANSWER: c.   12.97

36. How many moles of pure NaOH must be used to prepare 1.0 L of a solution that has pH = 12.00?
    a. 1.0×10^{-12} mol
    b. 0.010 mol
    c. 12.0 mol
    d. 1.0 mol
    e. none of these

    ANSWER: b.   0.010 mol

37. Calculate the pH of a 0.043 M HCl solution.
    a. 1.00
    b. 1.37
    c. 13.00
    d. 12.63
    e. none of these

    ANSWER: b.   1.37

38. Which statement is true for a strong base solution with a concentration greater than 1.0 M?
    a. pOH > pH
    b. pH > pOH
    c. pH < 0
    d. pH<14
    e. Two of these (a-d) are true.

    ANSWER: b.   pH > pOH

39. A solution is prepared by dissolving 100.0 g HCl(g) in enough water to make 150.0 L of solution. The pH of this solution is
    a. -0.176
    b. 1.74
    c. 4.25
    d. 9.75
    e. none of these

    ANSWER: b.   1.74

40. Calculate the pH of an acid solution containing 0.050 M HNO_3.
    a. 0.050
    b. 1.00
    c. 1.30
    d. 7.00
    e. none of these

    ANSWER: c.   1.30

41. Calculate the pH of 1.0×10^{-4} M HCl.
    a. 1.00
    b. 2.50
    c. 4.00
    d. 7.00
    e. none of these

    ANSWER: c.   4.00

42. Calculate the pH of 0.040 M $HClO_4$.
    a. 0.040
    b. 1.00
    c. 1.40
    d. 7.00
    e. none of these

    ANSWER: c.   1.40

43. What is the pH of a 2.0 M solution of HNO_3?
    a. -0.30
    b. 0.30
    c. 1.00
    d. 1.70
    e. none of these

    ANSWER: a.   -0.30

44. What is the pH of a 0.20 M nitric acid (HNO_3) solution?
    a. 0.70
    b. 2.00
    c. 10.15
    d. 13.30
    e. none of these

    ANSWER: a.   0.70

45. What is the pH of a 2.0 M solution of $HClO_4$?
    a. 13.70
    b. 0.30
    c. -0.30
    d. 14.30
    e. minus infinity

    ANSWER: c.   -0.30

46. Which of the following is *true* for a buffered solution?
    a. The solution resists any change in its $[H^+]$.
    b. The solution will not change its pH very much even if a concentrated acid is added.
    c. The solution will not change its pH very much even if a strong base is added.
    d. Any H^+ ions added will react with a conjugate base of a weak acid already in solution.
    e. all of these

    ANSWER: e.   all of these

47. A weak acid, HF, is in solution with dissolved sodium fluoride, NaF. If HCl is added, which ion will react with the extra hydrogen ions from the HCl to keep the pH from changing?
    a. OH^-
    b. Na^+
    c. F^-
    d. Na^+
    e. none of these

    ANSWER: c. F^-

48. A buffered solution contains $HC_2H_3O_2$. It also contains
    a. NaCl(aq)
    b. $H_2C_2H_3O_2(aq)$
    c. $NaC_2H_3O_2(aq)$
    d. KOH(aq)
    e. NaOH(aq)

    ANSWER: c. $NaC_2H_3O_2(aq)$

49. A NaCl(aq) solution is automatically buffered. T/F _____

    ANSWER: False

CHAPTER 16

Equilibrium

1. Chemists believe that chemical reactions occur because the molecules involved in the reaction
   _____.
   a. spontaneously break apart then recombine
   b. are always unstable
   c. exist only below a certain maximum temperature
   d. collide with each other with enough energy to break chemical bonds
   e. are moving so fast that the chance of interaction is very small

   ANSWER: d.   collide with each other with enough energy to break chemical bonds

2. Which of the following statements is(are) typically true for a catalyst?
   a. The concentration of the catalyst will decrease as a reaction proceeds.
   b. The catalyst provides a new pathway for the reaction.
   c. The catalyst speeds up the reaction.
   d. two of these
   e. none of these

   ANSWER: d.   two of these

3. True or false: The rate of a chemical reaction is directly related to the number of collisions that
   occur between reactant molecules.

   ANSWER: False

4. How many of the following are true?
   I.   At equilibrium, the concentrations of all reactants and products are equal.
   II. At equilibrium, all products and reactants coexist.
   III. At equilibrium the change in concentration over time for all reactants and products is
        zero.
   IV. At equilibrium, the rates of the forward and reverse reactions are zero.
   a. 0
   b. 1
   c. 2
   d. 3
   e. 4

   ANSWER: c.   2

5. Determine the equilibrium constant (units deleted) for the system $N_2O_4(g) \rightleftharpoons 2NO_2(g)$ at 25°C. The equilibrium concentrations are shown here:
$[N_2O_4] = 4.27 \times 10^{-2}\ M$ $[NO_2] = 1.41 \times 10^{-2}\ M$
   a. 0.330
   b. 3.03
   c. 0.660
   d. 215
   e. 0.00466

   ANSWER: e.   0.00466

6. Given the reaction $A(g) + B(g) \rightleftharpoons C(g) + D(g)$. You have the gases A, B, C, and D at equilibrium. Upon adding gas A, the value of K
   a. increases because by adding A, more products are made, increasing the product-reactant ratio
   b. decreases because A is a reactant so the product-to-reactant ratio decreases
   c. does not change because A does not figure into the product-to-reactant ratio
   d. does not change as long as the temperature is constant
   e. depends on whether the reaction is endothermic or exothermic

   ANSWER: d.   does not change as long as the temperature is constant

7. Write the equilibrium expression for the reaction
   $3O_2(g) \rightleftharpoons 2O_3(g)$

   ANSWER:
   $$K = \frac{[O_3]^2}{[O_2]^3}$$

8. For the reaction of gaseous nitrogen with gaseous hydrogen to produce gaseous ammonia, the correct equilibrium expression is

   a. $\dfrac{[NH_3]^2}{[N]^3[H_2]}$

   b. $\dfrac{[NH_3]}{[N_2][H_2]}$

   c. $\dfrac{[NH_3]^2}{[N_2][H_2]^3}$

   d. $\dfrac{[NH_3]}{[N][H]^3}$
   e.   none of these

   ANSWER: c. $\dfrac{[NH_3]^2}{[N_2][H_2]^3}$

9. The correct equilibrium expression for the reaction of sulfur dioxide gas with oxygen gas to produce sulfur trioxide gas is

   a. $\dfrac{[SO_3]}{[SO_2][O_2]}$

   b. $\dfrac{[SO_3]^2}{[SO_2]^2[O_2]}$

   c. $\dfrac{[SO_3]}{[SO_2]^2[O_2]}$

   d. $\dfrac{[O_2][SO_2]^2}{[SO_3]^2}$

   e. none of these

   ANSWER: b. $\dfrac{[SO_3]^2}{[SO_2]^2[O_2]}$

10. For the reaction
    $F_2(g) \rightarrow 2F(g)$
    at a particular temperature, the concentrations at equilibrium were observed to be $[F_2] = 1.0 \times 10^{-2}$ mol/L and $[F] = 2.0 \times 10^{-4}$ mol/L. Calculate the value of the equilibrium constant from these data. (The units are deleted.)
    a. 2.0×10^{-2}
    b. 2.0
    c. 2.5×10^5
    d. 4.0×10^{-6}
    e. none of these

    ANSWER: d. 4.0×10^{-6}

11. For the reaction
    $2NO(g) + O_2(g) \rightleftharpoons 2NO_2(g)$
    at a certain temperature, the equilibrium concentrations were found to be $[NO_2] = 3.0 \times 10^{-3}$ M, $[O_2] = 1.0 \times 10^{-2}$ M, and $[NO] = 2.0 \times 10^{-3}$ M. Calculate the value of the equilibrium constant from these data (delete units).
    a. 2.3×10^2
    b. 1.5×10^5
    c. 4.3×10^{-3}
    d. 2.3×10^{-2}
    e. none of these

    ANSWER: a. 2.3×10^2

12. Write the equilibrium expression for the following reaction:
    $S(s) + O_2 \rightleftharpoons SO_2(g)$

    a. $K = \dfrac{[SO_2]}{[O_2]}$

    b. $K = \dfrac{[O_2]}{[SO_2]^2}$

    c. $K = \dfrac{[SO_2]}{[O_2]^2}$

    d. $K = \dfrac{[SO_2]^2}{[O_2]}$

    e. none of these

    ANSWER: a. $K = \dfrac{[SO_2]}{[O_2]}$

13. Write the equilibrium expression for the following reaction:
    $CaCO_3(s) \rightleftharpoons CaO(s) + CO_2(g)$

    a. $K = [CaCO_3]$
    b. $K = [CO_2]$
    c. $K = [CaO]$
    d. $K = \dfrac{[CaO][CO_2]}{[CaCO_3]}$
    e. $K = \dfrac{[CaO]}{[CaCO_3]}$

    ANSWER: b. $K = [CO_2]$

14. Which of the following is an example of a homogeneous equilibrium?
    a. $MgCO_3(s) \rightleftharpoons MgO(s) + CO_2(g)$
    b. $NaCl(s) \rightleftharpoons Na^+(aq) + Cl^-(aq)$
    c. $3H_2(g) + N_2(g) \rightleftharpoons 2NH_3(g)$
    d. $C(s) + CO_2(g) \rightleftharpoons 2CO(g)$
    e. none of these

    ANSWER: c. $3H_2(g) + N_2(g) \rightleftharpoons 2NH_3(g)$

15. The equilibrium expression for the following reaction is
    $2A(s) + 3B(l) \rightleftharpoons C(aq) + 4D(aq)$

    a. $\dfrac{[A]^2[B]^3}{[C][D]^4}$

    b. $[C][D]^4$

    c. $\dfrac{[C][4D]^4}{[2A]^2[3B]^3}$

    d. $\dfrac{[C][D]^4}{[A]^2[B]^3}$

    e. none of these

    ANSWER: b. $[C][D]^4$

The following questions refer to the equilibrium shown here:
$CaCO_3(s) \rightleftharpoons CaO(s) + CO_2(g)$

16. What would happen to the system if more $CaCO_3$ were added?
    a. More CaO would be produced.
    b. The concentration of $CO_2(g)$ would decrease.
    c. The amount of $CaCO_3$ would decrease.
    d. The pressure would increase.
    e. Nothing would happen.

    ANSWER: e.   Nothing would happen

17. What would happen to the system if the total pressure were increased by adding $CO_2(g)$?
    a. Nothing would happen.
    b. More $CO_2(g)$ would be produced.
    c. The amount of CaO would increase.
    d. The amount of $CaCO_3$ would increase.
    e. Equilibrium would shift to the right.

    ANSWER: d.   The amount of $CaCO_3$ would increase.

18. What would happen to the system if the total pressure were increased by adding $Ar(g)$?
    a. Nothing would happen.
    b. More $CO_2(g)$ would be produced.
    c. The amount of CaO would increase.
    d. The amount of $CaCO_3$ would increase.
    e. Equilibrium would shift to the right.

    ANSWER: a.   Nothing would happen

Consider the reaction
$$2H_2(g) + O_2(g) \rightleftharpoons 2H_2O(g)$$
at some equilibrium position. Using the following choices, indicate what will happen if the following changes are made.
 a. shifts to the left
 b. shifts to the right
 c. no change

19. Additional $H_2O(g)$ is injected into the reaction vessel.

 ANSWER: a. shifts to the left

20. Some $H_2(g)$ is removed from the reaction vessel.

 ANSWER: a. shifts to the left

21. The size of the reaction vessel is decreased.

 ANSWER: b. shifts to the right

22. Some $He(g)$ is injected into the reaction vessel.

 ANSWER: c. no change

Consider the reaction system
$$CH_4(g) + 2O_2(g) \rightleftharpoons CO_2(g) + 2H_2O(g) + energy$$
and use the following choices to describe what happens when the following changes are made to the system at equilibrium.
 a. shifts to the left
 b. shifts to the right
 c. no change

23. $O_2(g)$ is removed from the reaction vessel.

 ANSWER: a. shifts to the left

24. $CO_2(g)$ is removed from the reaction vessel.

 ANSWER: b. shifts to the right

25. $He(g)$ is added to the reaction vessel.

 ANSWER: c. no change

26. The temperature is increased.

 ANSWER: a. shifts to the left

27. $CH_4(g)$ is added to the reaction vessel.

 ANSWER: b. shifts to the right

28. Consider the following equilibrium:
    $$2H_2(g) + X_2(g) \rightleftharpoons 2H_2X(g) + energy$$
    The equilibrium expression is

    a. $\dfrac{[2H_2X]}{[2H_2][X_2]}$

    b. $\dfrac{[2H_2X]}{[2H_2]+[X_2]}$

    c. $\dfrac{[H_2X]^2}{[H_2]^2+[X_2]}$

    d. $\dfrac{[H_2X]^2}{[H_2]^2[X_2]}$

    e. none of these

    ANSWER: d. $\dfrac{[H_2X]^2}{[H_2]^2[X_2]}$

29. Consider the following equilibrium:
    $$2H_2(g) + X_2(g) \rightleftharpoons 2H_2X(g) + energy$$
    Addition of X_2 to this system at equilibrium
    a. will cause $[H_2]$ to decrease
    b. will cause $[X_2]$ to increase
    c. will cause $[H_2X]$ to increase
    d. will have no effect
    e. cannot possibly be carried out

    ANSWER: a.   will cause $[H_2]$ to decrease

30. Consider the following equilibrium:
    $$2H_2(g) + X_2(g) \rightleftharpoons 2H_2X(g) + energy$$
    Addition of argon to this system at equilibrium
    a. will cause $[H_2]$ to decrease
    b. will cause $[X_2]$ to increase
    c. will cause $[H_2X]$ to increase
    d. will have no effect
    e. cannot possibly be carried out

    ANSWER: d.   will have no effect

222   *Chapter 16*

31. Consider the following equilibrium:
    $2H_2(g) + X_2(g) \rightleftharpoons 2H_2X(g) + energy$
    Decreasing the volume of the container for this system at equilibrium will cause
    a. an increase in the amount of H_2X
    b. an increase in the amounts of H_2 and X_2
    c. an increase in the amount of H_2 but not X_2
    d. no change
    e. X_2 to dissociate

    ANSWER: a.   an increase in the amount of H_2X

32. In the presence of ultraviolet light, the 'inert' gas xenon (Xe) will react with fluorine (F_2) gas to produce solid XeF_4. What is the equilibrium expression for this reaction?

    a. $\dfrac{[XeF_4]}{[Xe][F_2]}$

    b. $\dfrac{[XeF_4]}{[Xe][F_2]^2}$

    c. $\dfrac{1}{[F_2]}$

    d. $\dfrac{1}{[Xe][F_2]^2}$

    e. $\dfrac{1}{[Xe][F_2]}$

    ANSWER: d. $\dfrac{1}{[Xe][F_2]^2}$

Copyright © Houghton Mifflin Company. All rights reserved.

33. Consider the following equilibrium:
$H_2(g) + I_2(s) \rightleftharpoons 2HI(g)$
The equilibrium expression is

a. $\dfrac{[H_2][I_2]}{[HI]^2}$

b. $\dfrac{\sqrt{[H_2][I_2]}}{[HI]^2}$

c. $\dfrac{[HI]}{\sqrt{[H_2]}}$

d. $\dfrac{[HI]^2}{[H_2][I_2]}$

e. $\dfrac{[HI]^2}{[H_2]}$

ANSWER: e. $\dfrac{[HI]^2}{[H_2]}$

34. For a certain reaction at 25°C, the value of K is 1.2×10^{-3}. At 50°C, the value of K is 3.4×10^{-1}. This means that the reaction is
a. exothermic
b. endothermic
c. More information is needed.

ANSWER: b.   endothermic

Given the equation $A(g) \rightleftharpoons B(g) + 2C(g)$. At a particular temperature, $K = 1.4 \times 10^5$.

35. If you mixed 1.2 mol B, 0.050 mol C, and 0.003 mol A in a 1-L container, in which direction would the reaction initially proceed?
a. to the left
b. to the right
c. The mixture is in the equilibrium state.
d. cannot tell from the information given

ANSWER: b.   to the right

36. Addition of chemical B to an equilibrium mixture of the above
a. will cause [A] to increase
b. will cause [C] to increase
c. will have no effect
d. cannot be determined
e. none of the above

ANSWER: a.   will cause [A] to increase

37. Placing the equilibrium mixture in an ice bath (thus lowering the temperature)
    a. will cause [A] to increase
    b. will cause [B] to increase
    c. will have no effect
    d. cannot be determined
    e. none of the above

    ANSWER: d.   cannot be determined

38. Raising the pressure by decreasing the volume of the container
    a. will cause [A] to increase
    b. will cause [B] to increase
    c. will have no effect
    d. cannot be determined
    e. none of the above

    ANSWER: a.   will cause [A] to increase

39. Which of the following is true when the equilibrium constant for a reaction is relatively small?
    a. It will take a short time to reach equilibrium.
    b. It will take a long time to reach equilibrium.
    c. The equilibrium lies to the left.
    d. The equilibrium lies to the right.
    e. two of these

    ANSWER: c.   The equilibrium lies to the left.

Consider a system of four gases. The equilibrium concentration of each product is 1.3 M. The equilibrium concentrations of the reactants are equal. The equilibrium is shown here:
$$A + B \rightleftharpoons C + D \qquad K = 2.6$$

40. What is the equilibrium concentration of gas A?
    a. 2.5 M
    b. 0.20 M
    c. 1.2 M
    d. 0.80 M
    e. 1.6 M

    ANSWER: d.   0.80 M

41. For the reaction
    $$2SO_2(g) + O_2(g) \rightleftharpoons 2SO_3(g)$$
    at a certain temperature, the equilibrium concentrations were observed to be [SO_2] = 0.75 M, [O_2] = 5.5 x 10^{-2} M, and [SO_3] = 0.50 M.
    a. Calculate the value of K for this system at this temperature.
    b. In another experiment at the same temperature the equilibrium concentrations of SO_2 and SO_3 were found to be 1.5 M and 0.25 M, respectively. Calculate the equilibrium concentration of O_2.

    ANSWER:
    a. $K = 8.1$
    b. [O_2] = 3.4 x 10^{-3} M

42. For the reaction
$$CO(g) + H_2O(g) \rightleftharpoons CO_2(g) + H_2(g)$$
$K = 3.88$ at a certain temperature. If at this temperature in a certain experiment $[H_2] = 1.4\,M$, $[CO_2] = 1.8\,M$, and $[H_2O] = .26\,M$, calculate $[CO]$.
   a. $3.9\,M$
   b. $0.40\,M$
   c. $0.66\,M$
   d. $2.5\,M$
   e. none of these

   ANSWER: d. $2.5\,M$

43. Write the balanced equation for the dissolving of $PbCl_2(s)$ in water.

   ANSWER: $PbCl_2(s) \rightleftharpoons Pb^{2+}(aq) + Cl^-(aq)$

44. Write the balanced equation for the dissolving of $Ag_2S(s)$ in water.

   ANSWER: $Ag_2S(s) \rightleftharpoons 2Ag^+(aq) + S^{2-}(aq)$

45. Write the balanced equation for the dissolving of $Fe_3(PO_4)_2$ in water.

   ANSWER: $Fe_3(PO_4)_2(s) \rightleftharpoons 3Fe^{2+}(aq) + 2PO_4^{3-}(aq)$

46. The solubility of $BaCO_3(s)$ in water at 25°C is 4.0×10^{-5} mol/L. Calculate the value of the K_{sp} for $BaCO_3(s)$ at 25°C.
   a. 4.0×10^{-5}
   b. 6.3×10^{-3}
   c. 1.6×10^{-9}
   d. 8.0×10^{-5}
   e. none of these

   ANSWER: c. 1.6×10^{-9}

47. The solubility of $Mg(OH)_2(s)$ in water at 25°C is 1.3×10^{-4} mol/L. Calculate the K_{sp} for $Mg(OH)_2(s)$ at 25°C.
   a. 1.3×10^{-4}
   b. 1.7×10^{-8}
   c. 2.2×10^{-12}
   d. 8.8×10^{-12}
   e. none of these

   ANSWER: d. 8.8×10^{-12}

48. The K_{sp} for ZnS(s) is 2.5×10^{-22} at 25°C. The solubility of ZnS(s) in water at 25°C is
    a. 2.5×10^{-22} mol/L
    b. 1.6×10^{-11} mol/L
    c. 6.3×10^{-44} mol/L
    d. 1.3×10^{-22} mol/L
    e. none of these

    ANSWER: b. 1.6×10^{-11} mol/L

49. Given the solubility products (K_{sp})

$BaSO_4$	1.5×10^{-9}
CoS	5.0×10^{-22}
$PbSO_4$	1.3×10^{-2}
AgBr	5.0×10^{-13}

    which of the following compounds is the most soluble (in mol/L)?
    a. $BaSO_4$
    b. CoS
    c. $PbSO_4$
    d. AgBr
    e. $BaCO_3$

    ANSWER: c. $PbSO_4$

50. The solubility of $Fe(OH)_2$ in water is 7.9×10^{-6} mol/L at 25°C. The value of the K_{sp} of $Fe(OH)_2$ at 25°C is
    a. 4.9×10^{-16}
    b. 2.0×10^{-15}
    c. 6.2×10^{-11}
    d. 2.5×10^{-10}
    e. none of these

    ANSWER: b. 2.0×10^{-15}

51. The molar solubility of PbI_2 is 1.5×10^{-3} *M*. Calculate the value of K_{sp} for PbI_2.
    a. 1.5×10^{-3}
    b. 2.3×10^{-6}
    c. 1.4×10^{-8}
    d. 3.4×10^{-3}
    e. none of these

    ANSWER: c. 1.4×10^{-2}

52. Calculate the concentration of the silver ion in a saturated solution of silver chloride, AgCl (K_{sp} = 1.6×10^{-10}).
   a. 5.4×10^{-4}
   b. 1.3×10^{-5}
   c. 1.6×10^{-10}
   d. 8.0×10^{-11}
   e. none of these

   ANSWER: b. 1.3×10^{-5}

53. The concentration of Ag^+ in a saturated solution of Ag_2CrO_4 is 1.6×10^{-4} M. What is the K_{sp} value for Ag_2CrO_4?
   a. 2.0×10^{-12}
   b. 2.6×10^{-8}
   c. 4.1×10^{-12}
   d. 5.1×10^{-8}
   e. 1.6×10^{-11}

   ANSWER: a. 2.0×10^{-12}

54. The K_{sp} of Ag_2CO_3 is 8.1×10^{-12}. What is the molar solubility of silver carbonate?
   a. 1.3×10^{-4}
   b. 2.9×10^{-6}
   c. 2.0×10^{-4}
   d. 2.0×10^{-6}
   e. 8.2×10^{-12}

   ANSWER: a. 1.3×10^{-4}

55. The solubility in mol/L of Ag_2CrO_4 is 1.3×10^{-4} M at 25°C. What is the K_{sp} for this compound?
   a. 8.8×10^{-3}
   b. 6.1×10^{-9}
   c. 8.8×10^{-12}
   d. 4.7×10^{-13}
   e. 2.3×10^{-13}

   ANSWER: c. 8.8×10^{-12}

56. Silver chromate, Ag_2CrO_4, has a K_{sp} of 8.8×10^{-12}. What is the solubility in mol/L of silver chromate?
   a. 1.3×10^{-4} M
   b. 7.8×10^{-5} M
   c. 9.5×10^{-7} M
   d. 1.9×10^{-12} M
   e. 9.8×10^{-5} M

   ANSWER: a. 1.3×10^{-4} M

57. The solubility of $Cd(OH)_2$ in water is 1.7×10^{-5} mol/L at 25°C. The K_{sp} value for $Cd(OH)_2$ is

 a. 2.0×10^{-14}
 b. 4.9×10^{-15}
 c. 5.8×10^{-10}
 d. 2.9×10^{-10}
 e. none of these

 ANSWER: a. 2.0×10^{-14}

CHAPTER 17

Oxidation-Reduction Reactions and Electrochemistry

1. Consider the following reaction:
   $Ba(s) + F_2(g) \rightarrow BaF_2$
   Which of the following statements is *false*?
   a. The barium atom is gaining electrons; therefore, it is oxidized.
   b. The fluorine atom is gaining electrons; therefore, it is oxidized.
   c. The barium atom is losing electrons; therefore, it is oxidized.
   d. The fluorine atom is losing electrons; therefore, it is reduced.
   e. none of these

   ANSWER: c.   The barium atom is losing electrons; therefore, it is oxidized.

2. _____ is a loss of electrons.
   a. Reduction
   b. Neutralization
   c. Oxidation
   d. Galvanization
   e. None of these

   ANSWER: c.   Oxidation

3. In the reaction $2Ca(s) + O_2(g) \rightarrow 2CaO(s)$, calcium is _____.
   a. reduced
   b. electrolyzed
   c. synthesized
   d. oxidized
   e. none of these

   ANSWER: d.   oxidized

4. In the reaction $2Cs(s) + Cl_2(g) \rightarrow 2CsCl(s)$, the chlorine is _____.
   a. reduced
   b. oxidized
   c. synthesized
   d. electrolyzed
   e. none of these

   ANSWER: a.   reduced

5. The oxidation state of carbon in $NaHCO_3$ is
    a.   +6
    b.   +4
    c.   +2
    d.   0
    e.   -4

    ANSWER: b.   +4

6. The oxidation state of fluorine in NaF is
    a.   +1
    b.   0
    c.   -1
    d.   -3
    e.   -5

    ANSWER: c.   -1

7. The oxidation state of carbon in CO is
    a.   +6
    b.   +4
    c.   +2
    d.   0
    e.   none of these

    ANSWER: c.   +2

8. The oxidation state of sulfur in the sulfate ion is
    a.   +8
    b.   +6
    c.   +4
    d.   0
    e.   -2

    ANSWER: b.   +6

9. The oxidation state of chlorine in ClO_3^- is
    a.   +7
    b.   +5
    c.   +3
    d.   -3
    e.   -5

    ANSWER: b.   +5

10. The oxidation state of nitrogen in NO_2 is
    a.   +5
    b.   +4
    c.   +3
    d.   +1
    e.   none of these

    ANSWER: b.   +4

11. The oxidation state of phosphorus in PO_4^{3-} is
     a. -7
     b. -3
     c. +5
     d. +6
     e. +7

     ANSWER: c. +5

12. The oxidation state of Mg in any compound is
     a. +1
     b. +2
     c. +3
     d. -2
     e. -1

     ANSWER: b. +2

13. The oxidation state of nitrogen in N_2 is
     a. 0
     b. +1
     c. -2
     d. +8
     e. equal to its periodic table group number

     ANSWER: a. 0

14. In which of the following compounds does nitrogen have the most positive oxidation state?
     a. HNO_3
     b. NH_4Cl
     c. N_2O
     d. NO_2
     e. $NaNO_2$

     ANSWER: a. HNO_3

15. What is the oxidation state of Cl in $NaClO_3$?
     a. +1
     b. +3
     c. +5
     d. -1
     e. -7

     ANSWER: c. +5

16. What is the oxidation state of S in SO_3^{2-}?
    a. -2
    b. +6
    c. +4
    d. +2
    e. +8

    ANSWER: c.   +4

17. The oxidation state of Si in Na_2SiO_3 is
    a. +2
    b. -2
    c. -4
    d. +4
    e. +6

    ANSWER: d.   +4

18. The oxidation state of S in CaS_2O_3 is
    a. +2
    b. -2
    c. +4
    d. -4
    e. none of these

    ANSWER: a.   +2

19. The oxidation state of Mn in $NaMnO_4$ is
    a. +2
    b. +4
    c. +5
    d. +6
    e. +7

    ANSWER: e.   +7

20. Which of the following compounds contains chlorine with an oxidation state of +7?
    a. NaCl
    b. NaClO
    c. $NaClO_2$
    d. $NaClO_3$
    e. $NaClO_4$

    ANSWER: e. $NaClO_4$

21. The oxidation state of molybdenum in the polyatomic ion $Mo_2O_7^{2-}$ is
    a. +12
    b. -12
    c. +6
    d. -6
    e. -2

    ANSWER: c. +6

22. The oxidation state of chromium in $K_2Cr_2O_7$ is
    a. -2
    b. -6
    c. +6
    d. +7
    e. none of these

    ANSWER: c. +6

23. The oxidation state of C in Na_2CO_3 is
    a. +2
    b. -2
    c. +4
    d. -4
    e. +6

    ANSWER: c. +4

24. The oxidation state of Se in $CaSe_2O_3$ is
    a. +2
    b. -2
    c. +4
    d. -4
    e. none of these

    ANSWER: a. +2

25. The oxidation state of nitrogen in NH_4Cl is
    a. +1
    b. +3
    c. -3
    d. -4
    e. -2

    ANSWER: c. -3

26. The oxidation state of chlorine in Cl_2 is
    a. 0
    b. +1
    c. -1
    d. -2
    e. none of these

    ANSWER: a.   0

27. The oxidation state of carbon in CH_4 is
    a. +4
    b. -4
    c. +2
    d. -2
    e. none of these

    ANSWER: b.   -4

28. The oxidation state of carbon in CO_2 is
    a. +4
    b. -4
    c. +2
    d. -2
    e. none of these

    ANSWER: a.   +4

29. In reacting, metal atoms characteristically
    a. become anions
    b. lose electrons
    c. decrease in oxidation state
    d. form covalent bonds

    ANSWER: b.   lose electrons

30. Which statement is true of the following reaction?
    $2K + CuCl_2 \rightarrow 2KCl + Cu$
    a. Copper is reduced.
    b. Chlorine is reduced.
    c. Potassium is reduced.
    d. Copper is oxidized.
    e. Chlorine is oxidized.

    ANSWER: a.   Copper is reduced.

31. In which of the following does nitrogen have an oxidation state of +4?
    a. HNO_3
    b. NO_2
    c. N_2O
    d. NH_4Cl
    e. $NaNO_2$

    ANSWER: b. NO_2

32. The oxidation state of chromium in K_2CrO_4 is
    a. -1
    b. +3
    c. +7
    d. +8
    e. none of these

    ANSWER: e.   none of these

33. The following reaction occurs in aqueous acid solution:
    $NO_3^- + I^- \rightarrow IO_3^- + NO_2$
    The oxidation state of iodine in IO_3^- is
    a. 0
    b. +3
    c. -3
    d. +5
    e. -5

    ANSWER: d.   +5

Answer the questions that refer to the following reaction:
    $TiCl_4(l) + O_2(g) \rightarrow TiO_2(s) + 2Cl_2(g)$

34. Which species is oxidized?
    a. Ti
    b. Cl
    c. O
    d. TiO_2
    e. O_2

    ANSWER: b.   Cl

35. Which species is the reducing agent?
    a. Ti
    b. Cl
    c. O
    d. $TiCl_4$
    e. O_2

    ANSWER: d. $TiCl_4$

Answer the questions that refer to the following reaction:
$$SiO_2(s) + 2C(s) \rightarrow Si(s) + 2CO(g)$$

36. Which species is reduced?
    a. Si
    b. C
    c. O
    d. CO
    e. SiO_2

    ANSWER: a. Si

37. Which species is the oxidizing agent?
    a. Si
    b. C
    c. O
    d. SiO_2
    e. CO

    ANSWER: d. SiO_2

38. In the reaction $N_2(g) + 3H_2(g) \rightarrow 2NH_3(g)$, nitrogen is _____.
    a. oxidized
    b. reduced
    c. synthesized
    d. electrolyzed
    e. none of these

    ANSWER: b. reduced

39. In the reaction $P_4(s) + 10Cl_2(g) \rightarrow 4PCl_5(s)$, phosphorus is _____.
    a. reduced
    b. electrolyzed
    c. synthesized
    d. oxidized
    e. none of these

    ANSWER: d. oxidized

40. In the reaction $P_4(s) + 10Cl_2(g) \rightarrow 4PCl_5(s)$, chlorine is _____.
    a. oxidized
    b. reduced
    c. synthesized
    d. electrolyzed
    e. none of these

    ANSWER: b. reduced

41. In the reaction $C(s) + O_2(g) \rightarrow CO_2(g)$, carbon is _____.
    a. reduced
    b. electrolyzed
    c. synthesized
    d. oxidized
    e. none of these

    ANSWER: d. oxidized

42. Which of the following reactions does *not* involve oxidation-reduction?
    a. $CH_4 + 3O_2 \rightarrow 2H_2O + CO_4$
    b. $Zn + 2HCl \rightarrow ZnCl_2 + H_2$
    c. $2Na + 2H_2O \rightarrow 2NaOH + H_2$
    d. $MnO_2 + 4HCl \rightarrow Cl_2 + 2H_2O + MnCl_2$
    e. All are oxidation-reduction reactions.

    ANSWER: e.   All are oxidation-reduction reactions.

43. Which of the following are oxidation-reduction reactions?
    I. $PCl_3 + Cl_2 \rightarrow PCl_5$
    II. $Cu + 2AgNO_3 \rightarrow Cu(NO_3)_2 + 2Ag$
    III. $CO_2 + 2LiOH \rightarrow Li_2CO_3 + H_2O$
    IV. $FeCl_2 + 2NaOH \rightarrow Fe(OH)_2 + 2NaCl$
    a. III
    b. IV
    c. I and II
    d. I, II, and III
    e. I, II, III, and IV

    ANSWER: c. I and II

44. Which of the following statements is(are) true? Oxidation and reduction
    a. cannot occur independently of each other
    b. accompany all chemical changes
    c. describe the loss and gain of electron(s), respectively
    d. result in a change in the oxidation states of the species involved
    e. a, c, and d are true.

    ANSWER: e. a, c, and d are true.

45. In the reaction $Zn + H_2SO_4 \rightarrow ZnSO_4 + H_2$, which element, if any, is oxidized?
    a. zinc
    b. hydrogen
    c. sulfur
    d. oxygen
    e. None is oxidized.

    ANSWER: a. zinc

46. In the following reaction, which element is oxidized?
    $8NaI + 5H_2SO_4 \rightarrow 4I_2 + H_2S + 4Na_2SO_4 + 4H_2O$
    a. sodium
    b. iodine
    c. sulfur
    d. hydrogen
    e. oxygen

    ANSWER: b. iodine

47. Of the following four reactions, how many are oxidation-reduction reactions?
    $NaOH + HCl \rightarrow NaCl + H_2O$

    $Cu + 2AgNO_3 \rightarrow 2Ag + Cu(NO_3)_2$

    $Mg(OH)_2 \rightarrow MgO + H_2O$

    $N_2 + 3H_2 \rightarrow 2NH_3$
    a. 0
    b. 1
    c. 2
    d. 3
    e. 4

    ANSWER: c. 2

48. In the reaction shown below, which species is oxidized?
    $2NaI + Br_2 \rightarrow 2NaBr + I_2$

    a. Na^+
    b. I^-
    c. Br_2
    d. Br^-
    e. I_2

    ANSWER: b. I^-

49. In the following reaction, which species is the reducing agent?
    $3Cu + 6H^+ + 2HNO_3 \rightarrow 3Cu^{2+} + 2NO + 4H_2O$

    a. H^+
    b. Cu
    c. N in NO
    d. Cu^{2+}
    e. N in HNO_3

    ANSWER: b. Cu

Answer the following questions that refer to the unbalanced reaction shown below as it occurs in acidic solution:

_____ $Cr_2O_7^{2-}$ *(aq)* + _____ I^- *(aq)* → _____ Cr^{3+} *(aq)* + _____ I_2 *(s)*

50. Determine the coefficient for the iodide ions.
     a. 1
     b. 2
     c. 3
     d. 6
     e. 7

     ANSWER: d. 6

51. Determine the coefficient for the Cr^{3+} ions.
     a. 1
     b. 2
     c. 3
     d. 6
     e. 7

     ANSWER: b. 2

52. Determine the coefficient for water in the balanced equation for the reaction.
     a. 1
     b. 2
     c. 3
     d. 6
     e. 7

     ANSWER: e. 7

53. Balance the following half-reaction that occurs in acidic solution:
     MnO_4^- *(aq)* → Mn^{2+} *(aq)*

     ANSWER: MnO_4^- *(aq)* + 5e$^-$ + 8H$^+$ *(aq)* → Mn^{2+} *(aq)* + $4H_2O$*(l)*

54. Balance the following half-reaction that occurs in acidic solution:
     IO_3^- *(aq)* → I_3^- *(aq)*

     ANSWER: $3IO_3^-$ *(aq)* + 18H$^+$ *(aq)* + 16e$^-$ → I_3^- *(aq)* + $9H_2O$*(l)*

55. Balance the following half-reaction that occurs in acidic solution:
     SO_3^{2-} *(aq)* → SO_4^{2-} *(aq)*

     ANSWER: SO_3^{2-} *(aq)* + H_2O*(l)* → SO_4^{2-} *(aq)* + 2H$^+$ *(aq)* + 2e$^-$

56. Balance the following half-reaction that occurs in acidic solution:
     NO_3^- *(aq)* → NO_2*(g)*

     ANSWER: NO_3^- *(aq)* + 2H$^+$ *(aq)* + e$^-$ → NO_2*(g)* + H_2O*(l)*

57. Balance the following reaction that takes place in acidic solution:
    $Mg(s) + HCl(aq) \rightarrow MgCl_2(aq) + H_2(g)$

    ANSWER: $Mg(s) + 2HCl(aq) \rightarrow MgCl_2(aq) + H_2(g)$

58. Balance the following reaction that takes place in acidic solution:
    $Cu(s) + NO_3^-(aq) \rightarrow Cu^{2+}(aq) + NO(g)$

    ANSWER: $3Cu(s) + 2NO_3^-(aq) + 8H^+(aq) \rightarrow 3Cu^{2+}(aq) + 2NO(g) + 4H_2O(l)$

59. Balance the following reaction that takes place in acidic solution:
    $Mn(s) + NO_3^-(aq) \rightarrow Mn^{2+}(aq) + NO_2(g)$

    ANSWER: $Mn(s) + 4H^+(aq) + 2NO_3^-(aq) \rightarrow Mn^{2+}(aq) + 2NO_2(g) + 2H_2O(l)$

60. Balance the following reaction that takes place in acidic solution:
    $I^-(aq) + IO_3^-(aq) \rightarrow I_3^-(aq)$

    ANSWER: $6H^+(aq) + 8I^-(aq) + IO_3^-(aq) \rightarrow 3I_3^-(aq) + 3H_2O(l)$

61. For the reaction of sodium bromide with chlorine gas to form sodium chloride and bromine, the appropriation half-reactions are (ox = oxidation and re = reduction)
    a. ox: $Cl_2 + 2e^- \rightarrow 2Cl^-$ re: $2Br^- \rightarrow Br_2 + 2e^-$
    b. ox: $2Br^- \rightarrow Br_2 + 2e^-$ re: $Cl_2 + 2e^- \rightarrow 2Cl^-$
    c. ox: $Cl + + e^- \rightarrow Cl^-$ re: $Br \rightarrow Br^- + e^-$
    d. ox: $Br + 2e^- \rightarrow Br^-$ re: $2Cl^- \rightarrow Cl_2 + 2e^-$
    e. ox: $2Na^+ + 2e^- \rightarrow 2Na$ re: $2Cl^- \rightarrow Cl_2 + 2e^-$

    ANSWER: b. ox: $2Br^- \rightarrow Br_2 + 2e^-$ re: $Cl_2 + 2e^- \rightarrow 2Cl^-$

62. In the balanced equation for the following redox equation, the sum of the coefficients is
    $Fe^{3+} + I^- \rightarrow Fe^{2+} + I_2$
    a. 4
    b. 5
    c. 6
    d. 7
    e. 8

    ANSWER: d. 7

63. The following reaction occurs in aqueous acid solution:

    $NO_3^- + I^- \rightarrow IO_3^- + NO_2$

    In the balanced equation, the coefficient of water is
    a. 1
    b. 2
    c. 3
    d. 4
    e. 5

    ANSWER: c. 3

64. For the redox reaction $2Fe^{2+} + Cl_2 \rightarrow 2Fe^{3+} + 6Cl^-$, which of the following are the correct half-reactions?

        I. $Cl_2 + 2e^- \rightarrow 2Cl^-$

        II. $Cl \rightarrow Cl^- + e^-$

        III. $Cl_2 \rightarrow 2Cl^- + 2e^-$

        IV. $Fe^{2+} \rightarrow Fe^{3+} + e^-$

        V. $Fe^{2+} + e^- \rightarrow Fe^{3+}$
    a. I and IV
    b. I and V
    c. II and IV
    d. II and V
    e. III and IV

    ANSWER: a.   I and IV

65. When the following reaction is balanced in acidic solution, what is the coefficient of water?

    $Zn(s) + NO_3^-(aq) \rightarrow Zn^{2+} \cdot (aq) + NH_4^+ (aq)$
    a. 1
    b. 2
    c. 3
    d. 4
    e. none of these

    ANSWER: c. 3

66. In the balanced equation for the following reaction (in acidic solution)

    $ClO_3^- + Fe^{2+} \rightarrow Cl^- + Fe^{3+}$

    the coefficient of Fe^{2+} is
    a. 1
    b. 6
    c. 2
    d. 8
    e. 3

    ANSWER: b. 6

67. Which of the following is *true* for a galvanic cell based on the following reaction?
   $Zn(s) + Cu^{2+}(aq) \rightarrow Zn^{2+}(aq) + Cu(s)$
   a. The zinc is being reduced.
   b. The zinc serves as the anode.
   c. The Cu^{2+} ion is being reduced.
   d. The Cu serves as the anode.
   e. two of these

   ANSWER: e.   two of these

68. In a galvanic cell that employs the reaction
   $Cu(s) + 2Ag^+(aq) \rightarrow Cu^{2+}(aq) + 2Ag(s)$
   copper metal serves as the anode. T/F _____

   ANSWER: True

69. In a galvanic cell that employs the reaction
   $Pb(s) + 2H_2SO_4(aq) + PbO_2(s) \rightarrow 2PbSO_4(s) + 2H_2O(l)$
   lead metal serves as the cathode. T/F _____

   ANSWER: False

Answer the following questions for a galvanic cell that employs the reaction
   $2Na(l) + S(l) \rightarrow Na_2S(s)$

70. What element is oxidized?

   ANSWER: Na

71. What element is reduced?

   ANSWER: S

72. What element serves as the anode?

   ANSWER: Na

73. What element serves as the cathode?

   ANSWER: S

74. Which of the following battery types is commonly used in an automobile?
   a. dry cell
   b. alkaline dry cell
   c. lead storage
   d. mercury cell
   e. nickel-cadmium

   ANSWER: c.   lead storage

75. In a car battery, which of the following is the electrolyte?
    a. Pb(s)
    b. $PbO_2(s)$
    c. $PbSO_4(s)$
    d. $H_2SO_4(aq)$
    e. none of these

    ANSWER: d. $H_2SO_4(aq)$

76. In a lead storage battery, PbO_2 is found at the _____.
    a. cathode
    b. anode
    c. dry cell
    d. galvanic cell
    e. none of these

    ANSWER: a. cathode

77. The speed at which most metals oxidize in air is slower than expected because
    a. a thin layer of the metal oxide forms on the metal surface thus inhibiting corrosion
    b. cathodic protection naturally occurs
    c. corrosion only occurs under water
    d. the concentration of oxygen in the atmosphere is not high enough
    e. none of these

    ANSWER: a.   a thin layer of the metal oxide forms on the metal surface thus inhibiting corrosion

78. When a metal corrodes, what is happening chemically?
    a. The metal atoms lose electrons.
    b. The metal atoms gain electrons.
    c. Electrons are not involved.
    d. The metal is combining with nitrogen gas.
    e. none of these

    ANSWER: a.   The metal atoms lose electrons.

79. Aluminum resists corrosion in air because aluminum metal gains electrons rather than loses electrons. T/F _____

    ANSWER: False

80. Which of the following is *not* true regarding the term *electrolysis*?
    a. Electrolysis is a process where electrical energy is used to produce a chemical change.
    b. Electrolysis involves forcing a current through a cell to produce a change that would otherwise not occur.
    c. The electrolysis of water produces hydrogen gas and oxygen gas.
    d. Electrolysis is used to recharge a lead storage battery.
    e. All of these are true.

    ANSWER: e.   All of these are true.

81. The Hall process for making aluminum involves
    a. chemically reacting alumina with carbon
    b. electrolysis of an aqueous solution of Al_2O_3 and Na_3AlF_6
    c. electrolysis of a molten solution of Al_2O_3 and Na_3AlF_6
    d. none of the above

    ANSWER: c.   electrolysis of a molten solution of Al_2O_3 and Na_3AlF_6

CHAPTER 18

Radioactivity and Nuclear Energy

1. Which of the following is *not* an example of spontaneous radioactive process?
   a. alpha-decay
   b. beta-decay
   c. positron production
   d. autoionization
   e. electron capture

   ANSWER: d.   autoionization

2. If a nucleus captures an electron, describe how the atomic number will change.
   a. It will increase by one.
   b. It will decrease by one.
   c. It will not change because the electron has such a small mass.
   d. It will increase by two.
   e. It will decrease by two.

   ANSWER: b.   It will decrease by one.

3. Polonium is a naturally radioactive element decaying with the loss of an alpha particle.

   $$^{210}_{84}\text{Po} \rightarrow {}^{4}_{2}\text{H} + ?$$

   What is the second product of this decay?

   a. $^{214}_{86}\text{Rn}$

   b. $^{206}_{82}\text{Pb}$

   c. $^{206}_{85}\text{At}$

   d. $^{208}_{80}\text{Hg}$

   e. none of these

   ANSWER: b. $^{206}_{82}\text{Pb}$

4. Thorium-234 undergoes beta particle production. What is the other product?

    a. $^{234}_{91}$ Pa

    b. $^{234}_{89}$ Ac

    c. $^{233}_{90}$ Th

    d. $^{233}_{91}$ Th

    e.   none of these

    ANSWER: a. $^{234}_{91}$ Pa

5. Alpha particles are
    a.   electrons
    b.   protons
    c.   neutrons
    d.   helium nuclei
    e.   x-rays

    ANSWER: d.   helium nuclei

6. Choose the particle having a relative mass of 0 amu and a charge of 1+.
    a.   alpha particle
    b.   beta particle
    c.   proton
    d.   neutron
    e.   none of these

    ANSWER: e.   none of these

7. The atomic particle having a mass of 0 amu and a charge of 1- is
    a.   an electron
    b.   a neutron
    c.   an alpha particle
    d.   a proton
    e.   none of these

    ANSWER: a.   an electron

8. Complete the table for these radioactive particles:

Name	Charge	Mass Number
Alpha		
Beta		
Gamma		

    ANSWER:

Name	Charge	Mass Number
Alpha	2	4
Beta	-1	0
Gamma	0	0

9. Lithium-8 is a beta producer. The product nuclide is _____.

    ANSWER: $^{8}_{4}$Be

10. When $^{230}_{90}$Th decays by producing an alpha particle, the produce nuclide is _____.

    ANSWER: $^{226}_{88}$Ra

11. The element curium ($Z = 242$, $A = 96$) can be produced by positive-ion bombardment when an alpha particle collides with which of the following nuclei? Recall that a neutron is also a product of this bombardment.
    a. $^{249}_{98}$Cf
    b. $^{241}_{94}$Pu
    c. $^{241}_{95}$Am
    d. $^{239}_{92}$U
    e. $^{239}_{94}$Pu

    ANSWER: e. $^{239}_{94}$Pu

12. When $^{14}_{7}$N is bombarded by (and absorbs) a proton, a new nuclide is produced plus an alpha particle. The nuclide produced is _____.

    ANSWER: $^{11}_{6}$C

13. The cesium-131 nuclide has a half-life of 30 years. After 90 years, about 6 g remain. The original mass of the cesium-131 sample is closest to
    a. 30 g
    b. 40 g
    c. 50 g
    d. 60 g
    e. 70 g

    ANSWER: c.   50 g

14. The half-life of a radioactive nuclide is
    a. that period of time in which 25% of the original number of atoms undergoes radioactive decay.
    b. the time at which the isotope becomes nonradioactive
    c. that period of time in which 50% of the original number of atoms undergoes radioactive decay.
    d. the period of time it takes to reduce the radioactivity by 100%
    e. none of the above

    ANSWER: c.   that period of time in which 50% of the original number of atoms undergoes radioactive decay.

15. The iodine-131 nuclide has a half-life of 8 days. If you originally have a 625-g sample, after 2 months you will have approximately
    a. 40 g
    b. 20 g
    c. 10 g
    d. 5 g
    e. less than 1 g

    ANSWER: d.   5 g

16. A sample of a radioactive element decays to 6.25% of its original number of radioactive nuclides in 36 years. What is the half-life of this radioactive element?

    ANSWER: 9 years

17. A particular radioactive element has a half-life of 2.00 weeks. What percent of the original sample is left after 28.0 days?

    ANSWER: 25.0%

18. What radioactive nuclide is often used to date wooden artifacts?

    ANSWER: $^{14}_{6}C$

19. When the uranium-235 nucleus is struck with a neutron, the zinc-72 and samarium-160 nuclei are produced along with some neutrons. How many neutrons are produced?
    a. 2
    b. 3
    c. 4
    d. 5
    e. 6

    ANSWER: c.   4

When the uranium-235 nucleus is struck with a neutron, the cesium-144 and strontium-90 nuclei are produced along with some neutrons and electrons.

20. How many neutrons are produced?
    a. 2
    b. 3
    c. 4
    d. 5
    e. 6

    ANSWER: a. 2

21. How many electrons are produced?
    a. 2
    b. 3
    c. 4
    d. 5
    e. 6

    ANSWER: c. 4

22. When the palladium-106 nucleus is struck with an alpha particle, a proton is produced along with a new element. What is this new element?
    a. cadmium-112
    b. cadmium-109
    c. silver-108
    d. silver-109
    e. none of these

    ANSWER: d. silver-109

23. A nuclear reactor is really just a source of heat to change water to steam. T/F _____

    ANSWER: True

24. A breeder reactor is one that produces $^{238}_{92}$U from $^{235}_{92}$U. T/F _____

    ANSWER: False

25. Nuclear fusion uses heavy nuclides such as $^{235}_{92}$U as fuel. T/F _____

    ANSWER: False

26. Strontium-90 from radioactive fallout is a health threat because, like _____, it is incorporated into bone.
    a. iodine
    b. cesium
    c. iron
    d. calcium
    e. uranium

    ANSWER: d. calcium

CHAPTER 19
Organic Chemistry

1. True or false: In a molecule, carbon always forms bonds with four other elements.

   ANSWER: False

2. A double bond involves sharing _____ electrons.
   a. 2
   b. 3
   c. 4
   d. 5
   e. 6

   ANSWER: c. 4

3. Hydrocarbons always contain carbon and _____.

   ANSWER: hydrogen

4. Hydrocarbons are always saturated. T/F _____

   ANSWER: False

5. An unsaturated hydrocarbon must contain a _____ or _____ bond.

   ANSWER: double; triple

6. In a saturated hydrocarbon, the carbon-carbon bond angles are all _____.

   ANSWER: 109.5°

7. Write the formula for the saturated alkane that contains six carbon atoms.

   ANSWER: C_6H_{14}

8. Write the formula for the saturated alkane that contains eight carbon atoms.

   ANSWER: C_8H_{18}

9. Pentane has how many structural isomers?
   a. one
   b. two
   c. three
   d. four
   e. five

   ANSWER: c.   three

10. How many types of alkyl groups can arise from propane?
    a. none
    b. one
    c. two
    d. three
    e. four

    ANSWER: c.   two

11. A molecule whose name includes trimethyl contains _____ methyl substituents.

    ANSWER: three

12. Name the following molecule.

    $$CH_3CHCH_2CH_3$$
    $$|$$
    $$CH_3$$

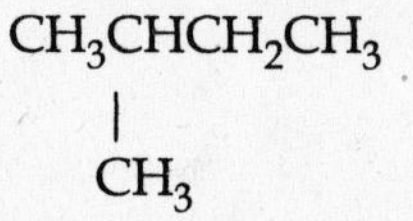

    ANSWER: 2-methylbutane

13. Name the following molecule.

    $$CH_3CHCH_2CHCH_3$$
    $$\quad\ |\qquad\quad |$$
    $$\quad\ CH_3\qquad CH_3$$

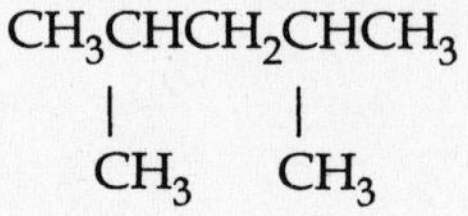

    ANSWER: 2,4-dimethylpentane

14. Name the following molecule.

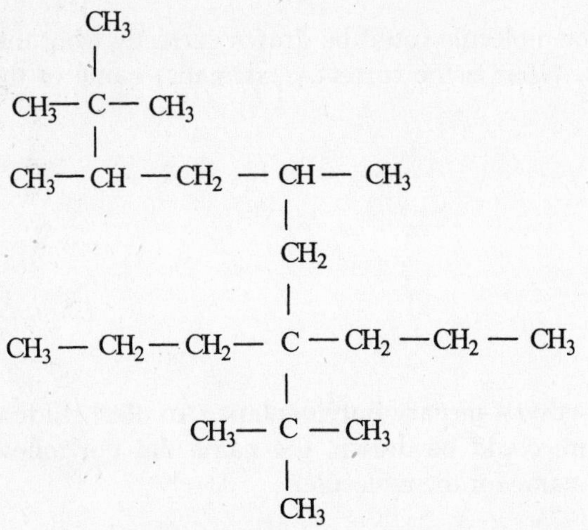

$$
\begin{array}{c}
CH_3 \\
| \\
CH_3-C-CH_3 \\
| \\
CH_3-CH-CH_2-CH-CH_3 \\
| \\
CH_2 \\
| \\
CH_3-CH_2-CH_2-C-CH_2-CH_2-CH_3 \\
| \\
CH_3-C-CH_3 \\
| \\
CH_3
\end{array}
$$

   a.   2,2,3,5-tetramethyl-7-propyl-7-*t*-butyldecane
   b.   6-propyl-2,6-di-*t*-butylnonane
   c,   2,2,5,7,8,8-hexamethyl-3,3-dipropylnonane
   d.   isonanane
   e.   none of these

   ANSWER: a.   2,2,3,5-tetramethyl-7-propyl-7-*t*-butyldecane

15. Name the following molecule.

$$
\begin{array}{c}
CH_3 \\
| \\
CH_3CH_2CHCH_2CH_3
\end{array}
$$

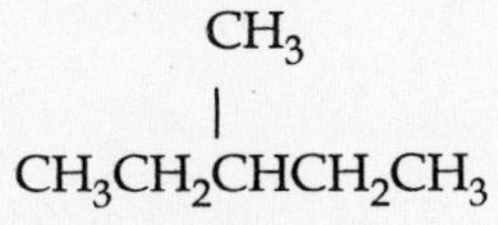

   a.   *n*-hexane
   b.   isohexane
   c.   1,2,3-trimethylpropane
   d.   methyl-diethylmethane
   e.   3-methylpentane

   ANSWER: e.   3-methylpentane

16. Name the following molecule.
   $CH_3 -- (CH_2)_7 - CH_3$

   a.   heptane
   b.   hexane
   c.   octane
   d.   nonane
   e.   decane

   ANSWER: d.   nonane

17. A student gave a molecule the following name:
2-ethyl-3-methyl-5-isopropylhexane
However, her TA pointed out that although the molecule could be drawn correctly from this name, the name violates the systematic rules. What is the correct (systematic) name of the molecule?
   a. 3,4-dimethyl-6-isopropylheptane
   b. 2-isopropyl-4,5-dimethylheptane
   c. 3,4,6,7-tetramethyloctane
   d. 1,2-diethyl-3,6,7-trimethylheptane
   e. 2,3,5,6-tetramethyloctane

   ANSWER: e.   2,3,5,6-tetramethyloctane

18. In lecture, the professor named a molecule 2-ethyl-4-tertiary-butylpentane. An alert student pointed out that although the correct structure could be drawn, the name did not follow systematic rules. What is the correct systematic name for the molecule?
   a. 2-*t*-butyl-5-methylhexane
   b. 2-ethyl-4,5,5-trimethylhexane
   c. 3,5,6,6-tetramethylheptane
   d. 2,2,3,5-tetramethylheptane
   e. undecane

   ANSWER: d.   2,2,3,5-tetramethylheptane

19.

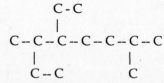

   is the carbon skeleton (minus any hydrogen atoms) of
   I. $C_{12}H_{26}$
   II.   a substituted octane
   III.   a compound with 3 tertiary carbons
   IV.   a compound with 3 secondary carbons
   V.   a compound with 2 isopropyl groups
   a. I, II, III
   b. II, III, IV
   c. III, IV, V
   d. II, IV, V
   e. I, II, III, IV

   ANSWER: a.   I, II, III

20.

$$
\begin{array}{ccccccc}
 & C & & & C & & \\
 & | & & & | & & \\
C\!-\!C\!-\!C\!-\!C\!-\!C\!-\!C\!-\!C \\
 & | & & | & & & \\
 & C\!-\!C & & C\!-\!C & & &
\end{array}
$$

is the carbon skeleton (minus any hydrogen atoms) of
a.   2,4-diethyl-3,6-dimethylheptane
b.   2,5-dimethyl-4,6-diethylheptane
c.   1,4-diethyl-3,6-dimethyl-tridecane
d.   5-ethyl-3,6-trimethyloctane
e.   4-ethyl-2,5,6-trimethyloctane

ANSWER: e.   4-ethyl-2,5,6-trimethyloctane

21.   Name the following molecule.

$$CH_3CHCH_2CHCH_3$$

$$
\begin{array}{cc}
| & | \\
CH_2 & CH_3 \\
| & \\
CH_3 &
\end{array}
$$

ANSWER: 2,4-dimethylhexane

22.   Draw the structural formula for 2,2-dimethylpropane.

ANSWER:

$$
\begin{array}{c}
CH_2 \\
| \\
CH_3\!-\!C\!-\!CH_3 \\
| \\
CH_3
\end{array}
$$

23.   Draw the structural formula for 3-ethyl-4-methylheptane.

ANSWER:

$$
\begin{array}{c}
CH_3 \\
| \\
CH_3CH_2CHCHCH_2CH_2CH_3 \\
| \\
CH_2 \\
| \\
CH_3
\end{array}
$$

24.   Gasoline consists of hydrocarbons with 20 to 30 carbon atoms. T/F _____

ANSWER: False

256 *Chapter 19*

25. Ethane undergoes dehydrogenation. The product of this undergoes an addition reaction with hydrogen gas. The product of this is
    a. ethylene
    b. propane
    c. butane
    d. ethane
    e. none of these

    ANSWER: d.   ethane

26. Which of the following cannot form alkenes?
    a. methane
    b. ethane
    c. propane
    d. two of the above
    e. all of the above

    ANSWER: a.   methane

27. Give the name of the molecule CH₃CH=CHCH₂CH₃.

    ANSWER: 2-pentene

28. Give the name of the molecule CH₃C=CHCH₃.

    ANSWER: 2-methyl-2-butene

29. The general name given to hydrocarbons with triple bonds is
    a. alkenes
    b. alkynes
    c. alkanes
    d. unsaturated hydrocarbons
    e. aromatic hydrocarbons

    ANSWER: b.   alkynes

30. CH₃C≡CCH₂CH₂Cl is named
    a. 1-chloro-3-pentyne
    b. 5-chloro-2-pentene
    c. 1-acetylenyl-3-chloropropane
    d. 5-chloro-2-pentyne
    e. 1-chloro-3-pentene

    ANSWER: d.   5-chloro-2-pentyne

31. Name the following molecule.

$$CH_2CH_3$$
$$|$$
$$CH_3\text{--}C\text{--}C \equiv C\text{--}H$$
$$|$$
$$H$$

   a. 1-hexyne
   b. 2-ethynyl butane
   c. 2-ethyl-3-butyne
   d. 3-methyl-1-pentyne
   e. 3-methyl-4-pentyne

ANSWER: d. 3-methyl-1-pentyne

32. Name the molecule $CH_3C \equiv CCH_2CH_3$.

ANSWER: 2-pentyne

33. Draw the structural formula for 3-methyl-2-hexene.

ANSWER:

$$CH_3CH=CCH_2CH_2CH_3$$
$$|$$
$$CH_3$$

34. Draw the structural formula for 4-ethyl-2-octyne.

ANSWER:

$$CH_3C \equiv CCHCH_2CH_2CH_2CH_3$$
$$|$$
$$CH_2$$
$$|$$
$$CH_3$$

35. Name the molecule below.

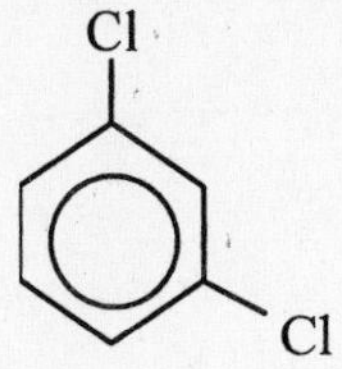

ANSWER: 1,3-dichlorobenzene   or   *m*-dichlorobenzene

36.   Draw the structural formula for 2-phenylpropane.

ANSWER:

CH₃CHCH₃

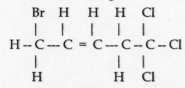

37.   Name the following:

Br   H   H   H   Cl
| | | | |
H -- C --- C = C --- C -- C -- Cl
| | |
H   H   Cl

a.   1,1,1-trichloro-5-bromo-3-pentene
b.   5,5,5-trichloro-1-bromo-2-pentene
c.   1,1,1-trichloro-5-bromo-2-pentene
d.   1,1,1-trichloro-5-bromo-3-pentyne

ANSWER: b.   5,5,5-trichloro-1-bromo-2-pentene

38.   Classify the following molecule.

H₃C -- C =O
|
H

a.   acid
b.   aldehyde
c.   amine
d.   ketone
e.   carbonyl

ANSWER: b.   aldehyde

39.   Which of the following has the greatest number of C -- O bonds?
a.   ketone
b.   ester
c.   alcohol
d.   amine
e.   aldehyde

ANSWER: b.   ester

40.   Which of the following has the least number of C -- O bonds?
a.   ketone
b.   alcohol
c.   ether
d.   ester
e.   Two of the above are equal.

ANSWER: b.   alcohol

41. Esters have sweet, fruity odors. T/F _____

    ANSWER: True

42. Give the name for the following molecule.

    CH$_3$CH$_2$CHCH$_3$
            |
           OH

    ANSWER: 2-butanol

43. Write the structural formula for 3-isopropyl-3-heptanol.

    ANSWER:

           OH
           |
    CH$_3$CH$_2$CCH$_2$CH$_2$CH$_2$CH$_3$
           |
         CH$_3$CHCH$_3$

44. Which of the following is known as wood alcohol?
    a. methanol
    b. ethanol
    c. propanol
    d. isopropanol
    e. none of these

    ANSWER: a.   methanol

45. Which of the following is known as rubbing alcohol?
    a. methanol
    b. ethanol
    c. propanol
    d. isopropanol
    e. none of these

    ANSWER: d.   isopropanol

46. Which of the following is found in beverages such as wine?
    a. methanol
    b. ethanol
    c. propanol
    d. isopropanol
    e. none of these

    ANSWER: b.   ethanol

47. Identify the type of organic compound shown.

    CH₃ -- C = O
              |
             CH₃

    a. aldehyde
    b. ester
    c. amine
    d. ketone
    e. none of these

    ANSWER: d. ketone

48. Identify the type of organic compound shown.

          H   H
          | |
    H --- C --- C --- C = O
          | | |
          H   H   H

    a. aldehyde
    b. ester
    c. amine
    d. ketone
    e. none of these

    ANSWER: a. aldehyde

49. Name the following compound.

    CH₃CH₂CH₂CHCH₂C = O
                | |
               Cl   H

    ANSWER: 3-chlorohexanal

50. Name the following compound.

          O
          ||
    H₃CHCH₂CH

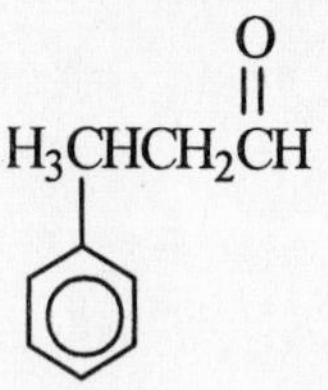

    ANSWER: 3-phenylbutanal

51. Draw the structural formula for pentanal.

    ANSWER:

    CH₃CH₂CH₂CH₂C = O
                    |
                    H

52. Draw the structural formula for 3,4-dichlorooctanal.

    ANSWER:
    $$CH_3CH_2CH_2CH_2CH\text{---}CHCH_2CH = O$$
    $$\quad\quad\quad\quad\quad\quad | \quad\quad |$$
    $$\quad\quad\quad\quad\quad\quad Cl \quad\quad Cl$$

53. Name the following compound.
    $$CH_3CCH_3$$
    $$\quad\quad ||$$
    $$\quad\quad O$$

    ANSWER: propanone (acetone)

54. The name methanal is the systematic name for
    a. acetone
    b. formaldehyde
    c. rubbing alcohol
    d. acetaldehyde
    e. water

    ANSWER: b.   formaldehyde

55. Name the following.
    $$\quad\quad\quad\quad Cl \quad\quad O$$
    $$\quad\quad\quad\quad | \quad\quad ||$$
    $$CH_3\text{---}CH\text{---}CH\text{---}C\text{---}CH(CH_3)_2$$
    $$\quad\quad\quad\quad |$$
    $$\quad\quad\quad\quad CH_2$$
    $$\quad\quad\quad\quad |$$
    $$\quad\quad\quad\quad CH_3$$

    a. 2-chloro-3-ethyl-1-isopropylbutanone
    b. isopropyl-chloromethylbutyl ketone
    c. 2-butylchloroisobutanoyl methane
    d. 4-chloro-2,5-dimethyl-3-heptanone
    e. 3-methyl-4-chloro-1-isopropylpentanone

    ANSWER: d.   4-chloro-2,5-dimethyl-3-heptanone

56. Name the following compound.
    $$CH_3COOH$$

    ANSWER: ethanoic acid (acetic acid)

57. Name the following compound.
    $$CH_3CHCH_2CCOOH$$
    $$\quad\quad | \quad\quad |$$
    $$\quad\quad Cl \quad\quad CH_3$$

    ANSWER: 4-chloro-2-methylpentanoic acid

58. Draw the structural formula for 3-ethylhexanoic acid.

ANSWER:

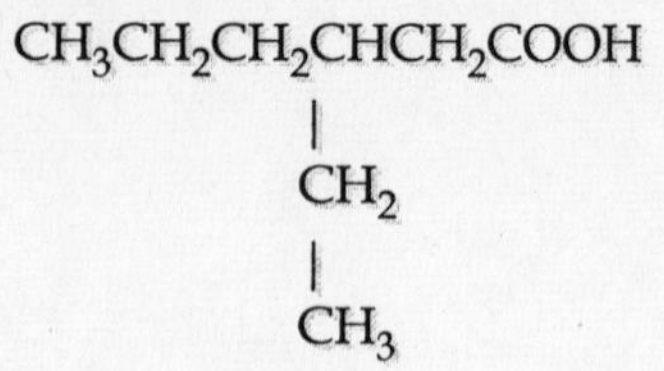

CH$_3$CH$_2$CH$_2$CHCH$_2$COOH
|
CH$_2$
|
CH$_3$

59. Nylon is an example of a
a. copolymer
b. homopolymer
c. dimer
d. two of these
e. none of these

ANSWER: a. copolymer

60. Teflon is an example of a
a. copolymer
b. homopolymer
c. dimer
d. two of these
e. none of these

ANSWER: b. homopolymer

CHAPTER 20

Biochemistry

1. How many of the following apply to fibrous proteins?
   I.   Provide structural integrity and strength for many types of tissues.
   II. Transport and store oxygen and nutrients.
   III. Act as catalysts.
   IV. Are the main components of muscle, hair, and cartilage.
   V.   Fight invasion of the body by foreign objects.
   a.   1
   b.   2
   c.   3
   d.   4
   e.   5

   ANSWER: b.   2

2. How many of the following apply to globular proteins?
   I.   Provide structural integrity and strength for many types of tissues.
   II. Transport and store oxygen and nutrients.
   III. Act as catalysts.
   IV. Are the main components of muscle, hair, and cartilage.
   V.   Fight invasion of the body by foreign objects.
   a.   1
   b.   2
   c.   3
   d.   4
   e.   5

   ANSWER: c.   3

3. The primary structure of proteins is
   a.   the pleated sheet
   b.   the peptide linkage
   c.   the alpha-helix
   d.   the order of amino acids
   e.   none of these

   ANSWER: d.   the order of amino acids

4. The condensation product of two amino acids is a(n)
    a. peptide
    b. ketone
    c. ether
    d. ester
    e. alcohol

    ANSWER: a. peptide

5. The primary structure of a protein refers to which amino acids are present in a protein. T/F
    _____

    ANSWER: False

6. Give one type of secondary protein structure.

    ANSWER: a-helix or pleated sheet

7. An example of a secondary structure of a protein is
    a. an alpha amino acid
    b. a peptide linkage
    c. a pleated sheet
    d. serine
    e. none of these

    ANSWER: c. a pleated sheet

8. The tertiary structure of a protein refers to how many carbons are bound to the a-carbon of the
    first amino acid. T/F _____

    ANSWER: False

9. Which statement is false with respect to proteins?
    a. Primary structure refers to the sequence of nucleotides.
    b. Secondary structure includes a-helixes.
    c. Tertiary structure includes disulfide bonds.
    d. The overall shape of a protein is related to the tertiary structure.
    e. All are false.

    ANSWER: a. Primary structure refers to the sequence of nucleotides.

10. The analysis of a protein for its amino acid content is valuable in determining the proteinÆs
    a. tertiary structure
    b. secondary structure
    c. quaternary structure
    d. primary structure

    ANSWER: d. primary structure

11. Denaturation of a protein means it is broken into its component amino acids. T/F _____

    ANSWER: False

12. The process of breaking down the three-dimensional structure of a protein is called
    a. degradation
    b. denaturation
    c. decomposition
    d. fission
    e. none of these

    ANSWER: b.   denaturation

13. Which of the following is *not* a function of proteins?
    a. structure
    b. catalysis
    c. oxygen transport
    d. energy transformation
    e. All of these are functions of proteins.

    ANSWER: e.   All of these are functions of proteins.

14. Biologic catalysts are called _____.

    ANSWER: enzymes

15. What protein carries oxygen from the lungs to the tissues?

    ANSWER: hemoglobin

16. A reaction being catalyzed occurs at the _____ of the enzyme.

    ANSWER: active site

17. The molecule that an enzyme acts on is called the _____.

    ANSWER: substrate

18. Simple sugars are more precisely called _____.

    ANSWER: monosaccharides

19. _____ is the major structural component of woody plants and natural fibers.

    ANSWER: Cellulose

20. Starch and cellulose are both polymers of glucose. T/F _____

    ANSWER: True

21. Name the three main generic components of DNA.

    ANSWER: deoxyribose; phosphate; organic base

22. What sugar is contained in RNA?

    ANSWER: ribose